External morphology and larval development of the Upper Cambrian maxillopod *Bredocaris admirabilis*

KLAUS J. MÜLLER and DIETER WALOSSEK

Müller, Klaus J. & Walossek, Dieter 1988 10 31: External morphology and larval development of the Upper Cambrian maxillopod *Bredocaris admirabilis*. *Fossils and Strata*, No. 23, pp. 1–70, Oslo. ISSN 0300-9491. ISBN 82-00-37412-2.

Search for new material yielded more than ninety specimens of different growth stages of *Bredocaris admirabilis* Müller, 1983. It enables us now to present an extended description of the largest stage, considered as adult, and of the larval sequence. Discovery of the tagma boundary of the cephalon behind the fifth pair of appendages led to the identification of the second maxilla, which has the same design as the thoracopods and is incorporated within the trunk limb series. The larval sequence comprises five successive metanaupliar instars, with delay of development of post-maxillulary limbs. Stages between larvae and the presumed adult have not been found. *Bredocaris*, about 0.85 mm long as adult, is assumed to have lived epibenthically, swimming closely above a flocculent bottom layer. Lack of special feeding structures on the trunk limbs and retention of the larval cephalic feeding apparatus in the adult suggest rather simple nutritory habits; filter feeding can be ruled out. The morphology, in particular the possession of seven pairs of thoracopods, and ontogeny indicate a systematic position of *Bredocaris* within the Maxillopoda and close alliance to the shield-bearing members of this subclass, the Thecostraca. Differences between *Bredocaris* and all known maxillopodan taxa is the basis for proposing the new order Orstenocarida and new family Bredocarididae. The major diagnostic characters of this new order include: a simple, posteriorly indented head shield, probably compound eyes, anterior three head appendages of naupliar shape, a 1st maxilla with rudimentary exopod and specialized for trophic function, a 2nd maxilla of trunk-limb shape, a thorax composed of seven segments, each with a pair of biramous paddle-shaped thoracopods, and a uniform abdomen carrying inarticulate, unsegmented furcal rami. □ *Crustacea, Maxillopoda, Orstenocarida*, Bredocaris, *cephalic shield, eyes, appendages, tagmosis, functional morphology, life habit, phylogeny of Crustacea, paedomorphosis.*

Klaus J. Müller and Dieter Walossek, Institut für Paläontologie, Rheinische Friedrich-Wilhelms-Universität, Nußallee 8, D-5300 Bonn 1, Federal Republic of Germany; 1988 02 03 (revised 1988 05 25).

Contents

List of abbreviations

Abd	abdomen
Ad	presumed adult
alp	alien particle
an	anus
app	appendages
atl	first antenna, antennula
a2	second antenna, antenna
bas	basipod
C	cephalic region, cephalon
cox	coxa
cs	cephalic shield
dcsp	dorsocaudal spine
den	denticles
do	'dorsal organ', plate-like structure on cephalic shield, with two pairs of pores at posterior margin
en	endopod
end	enditic process, endite
esp	enditic spine
ex	exopod
eye	probable compound eye
ff	frontal filaments of cirriped and facetotectan nauplii
fr	furcal ramus
fu	furrow
g	gut, digestive tract
gn	gnathobase, grinding plate of mandibular coxa
gns	gnathobasic seta, positioned on flattened distal surface of mandibular gnathobase
h	height
hol	holotype
i	initial, incipient
il	inner lamella, cuticle below shield
l	length
la	labrum
lo	lobes of probable compound eye
loc	locality
L1–5	larval stages
m	mouth
ma	area between lobes of probable compound eye and labrum
md	mandible
mx1	first maxilla, maxillula
mx2	second maxilla, maxilla
ne	naupliar eye
par	paratype
pce	proximal endite of post-mandibular appendages
pgn	paragnaths, pair of outgrowths on sternite of mandibular segment
prot	protopod
rud	rudimentary, of larval shape
s	seta
saf	supraanal flap, operculum
sh	shaft
spec	specimen
st	sternite, sternal bar
stl	setule, delicate thin bristle
T	trunk region of adults (= thorax + abdomen)
Th	thoracic region
thp1–7	thoracopods
tl	total length
tlb	trunk limb bud
tr	larval trunk, hind body
UB	repository number
w	width
wi	window, translucent cuticular area on dorsal shield of facetotectan nauplius, above naupliar eye

Introduction

From several *orsten* arthropods larval stages are known, such as from phosphatocopine ostracodes (Müller 1979, 1982b), the 'pre-crustacean' *Martinssonia elongata* (Müller & Walossek 1986a), and the agnostid trilobite *Agnostus pisiformis* (Müller & Walossek 1987). Various larvae are not assignable to any of the larger forms (Müller & Walossek 1986b; Walossek & Müller, in press). On the other hand, of the recently reviewed Skaracarida no larval stages could be encountered (Müller & Walossek 1985).

Whilst searching for more specimens of *Bredocaris admirabilis*, a large number of larval individuals were found, in some cases almost completely preserved, as well as additional specimens of the presumed adult stage. The material comes from eight samples of the Upper Cambrian zones 5 and 6 of the *orsten* sequence, from Västergötland, Sweden (productivity list in Table 1). As there are no significant morphological differences between the specimens from the different samples, the material is considered to be conspecific. The modes of processing and examination of these fragile phosphatized fossils have already been described in previous papers (e.g. Müller & Walossek 1985, pp. 3–4).

Of the 96 specimens (including the type material of Müller 1983) available for this study, 89 are assigned to a particular growth stage (Table 2; illustrated ones with repository number). Six different developmental stages were recognized on the basis of size measurements and morphology. The holotype (Müller 1983, Fig. 3A–C, UB 640) and the paratype (Müller 1983, Fig. 3D, UB 641) belong to the largest stage at hand, characterized by a set of well-developed trunk limbs. Its recognition as the adult stage by Müller (1983, p. 99) is supported herein. Approximate mean lengths of representative body parts (scheme of gross morphology in Fig. 1) from the different growth stages are given in Table 3.

Quality of preservation varies considerably among the material. Few specimens are more or less complete. In particular adult specimens are fragmentary, and commonly their head region is either distorted or not preserved at all (see Pls. 1–2). Many specimens are much wrinkled, ventral structures in particular being collapsed and shrunken (e.g. Pl. 7:5).

In order to better understand the ventral morphology, a model of the presumed adult, about 40 cm long (magnification ×500), was modelled with plasticine. It not only facilitated the reconstruction drawings but also helped in the interpretation of locomotion and feeding (Figs. 12–17). Except the mandibular coxa all appendages of the left side were omitted to permit a view of the median surface. The head shield was indicated by a wall only. Setation is almost complete, with the lengths of setae and spines approximated from the fossil material. Surface structures were simplified or omitted. Complete views of the animal were produced, when necessary, by combining a photograph with its mirror image.

The flexibility of the plasticine permitted study of the appendages in different positions. Therefore it was necessary to construct the protopods somewhat thinner than in the original, tolerating that the interlimb spaces are slightly enlarged. During the photographing, particularly the exopods and the setae became soft and tended to collapse from the heat of the photo lamps.

Taxonomic status

In the original description the second maxilla was interpreted as the first trunk limb (Müller 1983). Accordingly, the cephalon should have included only four limb-bearing segments, which contrasts with the definition of Crustacea. However, in two of the new specimens the tagma boundary between the head and thorax can be clearly recognized behind the 1st trunk-limb-like appendage, now identified as 2nd maxilla. With this correction, the cephalon includes five limb-bearing segments, as is typical of the Crustacea, and the thorax is composed of seven rather than eight limb-bearing segments.

Bredocaris is clearly a true crustacean, as previously suggested by Müller (1983). With respect to affinities, similarities to the Maxillopoda are greater than to any other subclass. Close alliance is apparent not only in gross morphology, in particular the tagmosis (see below), but also in larval morphology and mode of development. Several characters of *Bredocaris* are shared only with particular members of the Maxillopoda. These and the remarkable similarities to the 'podid' stages of this subclass (term introduced by Newman 1983 for the different post-naupliar stages, the copepodids and cyprids) give convincing evidence for a systematic position within the Maxillopoda.

On the other hand, *Bredocaris* lacks all major diagnostic features of the other subclasses, the Malacostraca, Branchiopoda, Cephalocarida, and Remipedia, which renders relationships with these unlikely. The first three groups show a completely different body segmentation, distinct modes of development, basically with large series of larval and post-larval instars, and different morphogenesis of structures. The recently discovered Remipedia (Yager 1981) possess more than 25 trunk segments, each carrying a pair of limbs, and specialized mouth parts, and grasping maxillipeds (Yager & Schram 1986; Schram, Yager, & Emerson 1986).

Table 1. Sample productivity (na = not assignable).

Sample	Ad	L1	L2	L3	L4	L5	na	total
975	–	–	1	–	1	1	–	3
5940	2	–	–	1	–	3	–	6
5948	5	3	5	–	2	2	1	18
5952	–	–	1	–	–	–	–	1
5955	2	–	–	–	1	–	–	3
5957	8	6	3	–	2	8	3	30
6392	2	–	1	–	–	–	–	3
6393	3	6	7	2	4	7	3	32
Total	22	15	18	3	10	21	7	96

The lack of detailed information on the Remipedia, in particular with regard to ontogeny and anatomy, presently precludes detailed comparison between this group and *Bredocaris*.

Exclusion of close affinities with *Bredocaris* does not, however, imply exclusion of similarities in particular of forms that have retained primordial features, such as certain primitive Malacostraca (e.g. euphausids, peneids; cf. Kaestner 1967; Mauchline 1971; Weigmann-Haass 1977; Cals 1978) or Cephalocarida (cf. Gooding 1963; Sanders 1963; Hessler 1964; Sanders & Hessler 1964; Hessler & Sanders 1971; Knox & Fenwick 1977; Burnett 1981). Among the Branchiopoda in particular the Devonian lipostracan *Lepidocaris rhyniensis* (Scourfield 1926, 1940) exhibits some similarity with *Bredocaris* in the anterior (=naupliar) body region and gross morphology of naupliar appendages, to a lesser degree also in the shape of post-mandibular limbs, furca, and pattern of setation and ornamentation with setules and denticles. Although probably plesiomorphic, these features may be useful for comparisons within a particular group or they may provide valuable information on the body plan of ancestral crustaceans, but they are of little value for determining the systematic relationships of *Bredocaris*.

Few Paleozoic crustaceans are preserved in a quality to provide information for detailed comparison with *Bredocaris*. Some *orsten* forms share the nauplius-like appearance of the anterior cephalic region with *Bredocaris*, but this similarity demonstrates only the retention of the basic design of the 'naupliar head region'. Besides superficial similarities, none of these forms shares characters with *Bredocaris* that indicate relationships, with two exceptions, the Skaracarida and probably *Dala peilertae* (see p. 30).

Two principal differences between *Bredocaris* and all known maxillopodan taxa are the presence of seven well-developed thoracopods and the trunk-limb shape of the second maxilla. It is generally accepted that the trunk of these groups is basically composed of six limb-bearing thoracic segments and an abdomen with five segments, the last including the telson (body plan: 5+6+5).

Grygier (1983a), however, predicted a seven-segmented thorax with an additional pair of limbs for the maxillopodan groundplan. This assumption was based on the presence of penis structures and on the seventh trunk segment in several Recent maxillopodan forms, which this author interpreted as modified limbs. In consequence, the first abdominal segment, in various groups bearing the genital openings, shifts onto the thorax, whereas the abdomen should have only four segments including the telson (body plan: 5+7+4). Evi-

Table 2. Reference list of examined and illustrated specimens.

Adults

UB	spec	sample	plate
882	1411	5948	1:1,2; 5:6
640	1417	5955	1:3, 4; 4:5–9; 5:1, 2; 6:1; 16:3 (holotype)
883	1400	5940	1:5; 3:6; 4:1
884	1419	5957	1:6; 6:2, 5
641	1418	5955	1:7; 3:1, 4, 5; 4:2, 3 (paratype)
885	1456	5957	1:8 (=2459)
886	1973	5948	1:9; 4:4; 5:3, 4; 16:2
887	1493	5957	2:1; 3:2; 6:3
888	2011	5957	2:2; 3:3
889	2109	6392	2:3; 16:1
890	1484	5957	2:4, 5
891	2066	6393	2:6; 6:8, 9
892	1454	5957	2:7
893	1492	5957	2:8
894	2470	6393	2:9; 5:5; 6:10; 16:4 (=2464)
895	1957	5948	3:7, 8
896	2085	6393	6:4, 7
897	2475	6392	6:6
932	6693	5940	16:5–8
	1458	5957	no
	1970	5948	no
	1971	5948	no

L1

UB	spec	sample	plate
898	1962	5948	7:1, 3, 8; 8:1, 3, 6–8
899	1965	5948	7:2, 6
900	2467	6393	7:4
901	1457	5957	7:5; 8:2, 5, 10
902	1489	5957	7:7; 8:11
903	2534	6393	7:9
904	2468	6393	8:4
905	2065	6393	8:9
	1422	5957	no
	1495	5957	no
	1862	5957	no
	1963	5948	no
	2024	5957	no
	2462	6393	no
	2472	6393	no

L2

UB	spec	sample	plate
906	2115	6393	9:7; 10:4, 5
907	1420	5957	9:2; 11:6
908	1413	5948	9:3, 6; 10:2; 11:5
909	2533	6393	9:4
910	1958	5948	9:5; 10:1; 11:4
911	1964	5948	9:1; 10:9
912	2096	6393	9:8, 9; 10:3, 6, 7; 11:1–3
913	2116	6393	10:8
914	2474	6393	10:10, 11
915	1734	975	11:7
916	1924	5952	11:8
	1487	5957	no
	1966	5948	no
	1968	5948	no
	1999	5957	no
	2060	6393	no
	2113	6392	no
	2461	6393	no

L3

UB	spec	sample	plate
917	2473	6393	1:9, 10
	1522	5940	no
	2100	6393	no

A

E

cox

bas

ex

Fig. 3.
antenn
and en

smooth
base (
also o
4:4, se
Fine
gnatho
be seer
adult,

L4

UB	spec	sample	plate
918	1412	5948	12:1, 2, 4–8; 13:1, 2, 4, 5, 8, 9, 11
919	1986	5955	12:3
920	2093	6393	13:3, 6, 10; 14:2
921	1520	5948	13:7
922	2103	6393	14:1
	1667	975	no
	2006	5957	no
	2019	5957	no
	2174	6393	no
	2471	6393	no

L5

UB	spec	sample	plate
923	1415	5940	14:3, 8; 15:1, 2, 6, 8
924	1509	5940	14:4; 15:4
925	1459	5957	14:5 (= 2460)
926	2082	6393	14:6
927	1453	5957	14:7
928	2466	6393	15:5
929	1414	5948	15:7
930	1466	5957	15:9
931	2008	5957	15:10
	1421	5957	no
	1464	5957	no
	1465	5957	no
	1469	5948	no
	1494	5957	no
	1524	5940	no
	1740	975	no
	2095	6393	no
	2101	6393	no
	2463	6393	no
	2464	6393	no
	2469	6393	no

not assignable: 1972–5948, 2025–5957, 2117–6393, 2119–6393, 1463–5957, 2522–6393.

dence for this conclusion of Grygier (1983a) may also be seen in the discovery of an ascothoracid larva with seven pairs of limb anlagen in a set underneath the cuticle (Grygier, personal communication) and probably also the illustrations of Walley (1969, in particular Fig. 5) of a late nauplius, showing a small hump behind the sixth thoracic limb bud.

The possession of seven well-defined thoracopods in *Bredocaris* fits closely with the suggested basic plan of the Maxillopoda. The apparent differences in detail, however, leads us to include this fossil form in a new order of Maxillopoda. Possible alliance to the shield-bearing core group around the Thecostraca and consequences are discussed below (pp. 26–30).

Class Crustacea Pennant, 1777

Subclass Maxillopoda Dahl, 1956 (*sensu lato*)

Order Orstenocarida n. ord.

Etymology of name. – After the local name of the limestones from which the present material was etched, and 'carida' meaning shrimps.

Composition. – One monogeneric family, Bredocarididae n. fam.

Geological range. – Known only from the Alum Shale, zones 5–6, Upper Cambrian of Sweden.

Diagnosis. – Tiny marine maxillopods with a simple cephalic shield, a nauplius-like anterior head, probably with compound eyes, large ventrally projecting labrum, and biramous mandibles with strong gnathobase, first maxillae with rudimentary exopod, second maxillae like trunk limbs, and a trunk divided in a feebly segmented thorax with seven pairs of biramous thoracopods and a uniform abdomen with inarticulate and unsegmented furcal rami. Larval sequence with five metanaupliar stages. Development delayed in all post-maxillulary limbs. Post-naupliar stages possibly absent, implying direct metamorphosis to adult.

Family Bredocarididae n. fam.

Etymology. – After the type genus.

Composition. – Monogeneric.

Diagnosis. – As for type species of *Bredocaris*.

Genus *Bredocaris* Müller, 1983

Type species. – Bredocaris admirabilis Müller, 1983.

Composition. – Monospecific.

Diagnosis. – As for the type species.

Bredocaris admirabilis Müller, 1983

Synonymy. – □ 1981a Metanaupliar larva of a crustacean – Müller, p. 150, Fig. 5. □ 1981b Undescribed small crustacean and a larva – Müller, pp. 14, 15. □ 1982a *Bredegcare admirabilis* Müller, 1982 [nom. nud.] – Müller, p. 252, Fig. 7. □ 1983 *Bredocaris admirabilis* n.sp. – Müller, pp. 97-99, Figs. 3, 4.

Material. – Holotype UB 640, paratype UB 641, 19 adult specimens, 67 larval specimens, and 6 unassignable ones.

Emended diagnosis. – Largest stage, presumably the adult, about 800–850 μm long, measured from tip of head to end of furcal rami. Body composed of three tagmata: cephalon with five pairs of appendages, thorax with seven limb-bearing segments, and unsegmented abdomen.

Cephalic shield univalve, roof-like, posteriorly indented. Outline suboval in dorsal view, widest in the first third. Probable compound eye consisting of two ovoid blisters, separated by a slightly bulging area. Labrum prominent and almost ventrally projecting. Shape subcylindrical and with rounded tip. Anterior cephalic region nauplius-like.

1st antenna 17-segmented. Distal portions of 2nd antenna and mandible similar, with prominent basipod and well-developed setation, four-segmented endopods, and multiannulated exopods. Mandibular coxa prominent, with strong

ed on all medially pointing parts of 2nd antenna, mandible and 1st maxilla. They occur in rows at the sides of the labrum, on the postoral sternal plate, and on the distal surface of the mandibular gnathobase. Setules are also present on the sides of the antennal exopod close to the basipod, as well as on the tips of 1st antenna, endopods of 2nd antenna and mandible, and exopod of 1st maxilla. In particular on the 1st maxilla the setules may convert into the shorter and more acute denticles towards the sides of the limb.

Setules obviously were elements of the feeding apparatus, either to brush off particles (when on movable parts) or to keep or guide food (see also Schrehardt 1986, Fig. 6, for *Artemia*). Their absence on the median margins of the postmaxillulary limbs reflects the function of these limbs (see p. 23).

(8). Denticles occur in rows on the annulations of the three anterior head appendages. Isolated ones or small groups can be seen on the sides of the limbs and their tips. The denticles occur in small, half-crescentic rows, commonly with a scale-like appearance, on the anterior surface of the labrum and the caudal end of the abdomen including the rami. The function of these denticles is unknown. It is possible that they had a hydrodynamic function, in order to support motion in an environment of low Reynolds number. Similar denticles are developed in various other fossil and Recent crustaceans and often in an identical position and pattern. This may indicate that they represent a very primitive surface structure of the Crustacea.

Pores. – Due to the exceptional mode of preservation, pores have been variously recognized in the *orsten* material. In *Bredocaris*, however, only a group of four pores could be found at the posterior margin of a plate-like structure on the apex of the cephalic shield, most likely present in all developmental stages (Pls. 3:2; 10:1).

Strikingly similar structures, which disappear during ontogeny, have been reported from certain larval euphausid and decapod Malacostraca (Mauchline 1971, 1977, in particular his Fig. 12; Laverack & Barrientos 1985; Barrientos & Laverack 1986). Equatorial pores are also present in larval Ascothoracida (Grygier 1984b). Their homologization is difficult, since such structures have not been studied in detail as yet. The 'dorsal organ' of the first nauplius of *Artemia*, which aids in osmoregulation (Schrehardt 1986, Fig. 2) and similar 'neck organs' of larvae of Conchostraca (e.g. Rieder *et al.* 1984) seem to be not homologous to that of other Crustacea. Similar dorsal pore groups are, however, not restricted to Crustacea but have been reported also from Trilobita (cf. Harrington 1959 for older references; Barrientos & Laverack 1986; Müller & Walossek 1987). Detailed comparisons, functional analysis, and an assessment of homology between the different types, are still lacking.

Infraspecific variation of morphological details. – On the basis of this material, individual variation appears to be very low, recognizable in the number and distribution pattern of setules and denticles, and in the expression of the larval limb buds.

Larval development

General remarks

The 67 larval specimens are grouped into five successive instars, based mainly on progressive increase in size (Table 3) and gradual increase in the number of limbs. All instars have in common the gross shape and details, such as the cephalic shield, eye, prominent labrum, the conical and somewhat ventrally oriented trunk, dorsocaudal spine, inarticulate incipient furcal rami, and the specific pattern of denticles on labrum and trunk. The last four instars and the presumed adult (see below) are linked with one another in particular by the functional 1st maxilla (see Figs. 4, 5).

Description of the stages

Instar I (L1; Pls. 7, 8; Fig. 4A). – Body length about 240 μm. Body divided into two distinct portions: anterior part with weakly defined shield and three pairs of appendages; hind body (tr) conical, with lobate rudimentary 1st maxilla anteroventrally, dorsocaudal spine at truncate end, terminal anus, and short incipient furcal rami (i fr).

Shield slightly arched and almost circular in outline, about 160 μm long, 150 μm wide, and 50 μm high. Anterior and posterior margins slightly raised towards the middle. Margins not well defined, shield only laterally slightly protruding from the body, in some specimens rather inconspicuous (in particular UB 1962, a well-inflated and complete specimen; Pl. 7:3, 8). Shield seems to extend post the maxillulary segment (Pl. 7:6).

Eye prominent, in some specimens protruding from the forehead (Pls. 7:2, 4–6, 8; 8:1, 2). Lobes oval to subtriangular, about 25 μm wide and 20 μm long, midventral area subtriangular, narrowing anteriorly and extending between the eye lobes. Area carrying a small knob or pore about 3–4 μm in diameter (Pl. 8:2).

Labrum 105 μm long, i.e. more than one third as long as the body, projecting almost perpendicularly from the head (Pl. 7). Shape loaf-like, with suboval cross-section, almost parallel-sided but broadly rounded tip. Anterior surface adorned with short rows of denticles (Pl. 7:5; Fig. 4A). Sides with two rows of posteriorly pointing setules running distally along labrum to about three fourth of its length. Mouth underneath labrum, deeply recessed (Pl. 8:7, 8). Mandibular sternite rectangular, wider than long, slightly raised but with depression in the middle, gently sloping towards the mouth. Surface covered with fine setules, arranged in symmetrical pattern of medially and orally directed rows (Pl. 8:7, 8).

Gross shape and insertion of 1st antenna as in adult, cross-section almost circular. 1st antenna about as long as the labrum, but due to its insertion not reaching to the tip of the latter (Pls. 7:5, 8; 8:1, 3–6). Anterior surface with 11–12 incomplete ringlets. Posterior surface pliable, as in adult, but only two setae medially (Pl. 8:5). Similar to adult, distal end made up from two cylindrical segments and a tiny hump forming the basis of the largest apical seta. Cylindrical segments with rows of denticles on outer surface, penultimate and terminal segments with additional setules around the setae.

Second antenna essentially as in adult, but shorter: about 160 μm long from proximal joint to tip of endopod, and 175

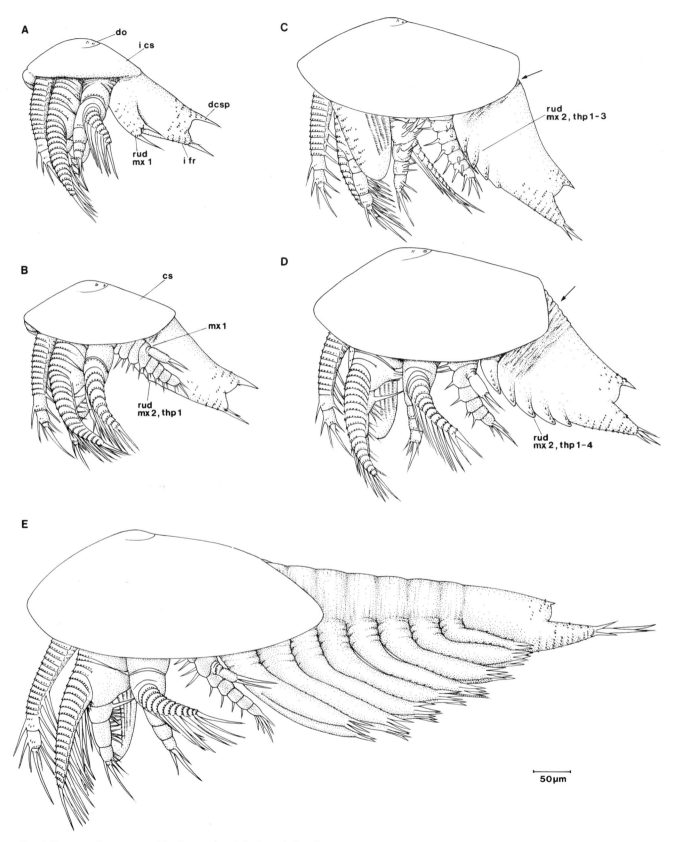

Fig. 4. Ontogenetic sequence of *Bredocaris admirabilis*, lateral view, but stages drawn in slightly different positions. □A. First metanauplius (L1). □B. Second metanauplius (L2). □C. Fourth metanauplius (L4), arrow points to developing pliable dorsal anterior trunk region. □D. Fifth metanauplius (arrow as in preceding figure); □E. Adult stage.

µm to tip of exopod (Pls. 7:3, 4, 8; 8:3, 5, 6). Exopod with 14–15 ringlets but only nine to ten setae, because in the middle of the ramus the setal sockets are thicker than the annuli. Coxa, basipod, and proximal endopodal podomere with robust enditic spines flanked by thinner setae (Pls. 7:1; 8:3, 7).

Mandible similar to adult one, shorter than 2nd antenna: length to tip of endopod 125 µm, and 100 µm to tip of exopod. Coxal gnathobase being a slightly flattened, pointed protrusion, tipped by a small spinule and slightly angled against the coxal body; margin posterior to spinule with about four small spinules, posterior one somewhat separated; gnathobasic seta present; gnathobase apparently less shovel-like than in later stages (Pl. 8:7, 8). Endopod four-segmented; exopod with eight to nine ringlets and seven setae (Pls. 7:7; 8:3, 6).

Trunk about 115 µm long, separated from anterior body portion by a deep transverse incision behind the sternal plate of the mandibular segment and mostly somewhat ventrally bent. Trunk tapering conically toward its truncate end. Maximum width and height about 75 µm (Pls. 7:1–7; 9).

First maxilla present as lobate straight posteriorly projecting protrusion, lying on ventral body wall, 40–50 µm long, bifid, inner outgrowth distinctly larger. Both ends, representing incipient endo- and exopods, tipped with a spinule. Distal surface furnished with few denticles (Pls. 7:1–7, 9; 8:6, 7, 10; Figs. 4A, 5A).

Anus located as in adult. Dorsocaudal spine prominent and robust, about 40 µm long. Incipient furcal rami cone-like, very short, about 15 µm long, forming the ventrocaudal extension of the trunk. Terminal end of rami tipped by two setae, a thicker median one and a thin, shorter one laterally. Surfaces of caudal part of trunk and furcal rami furnished with numerous groups and single denticles (Pl. 7:1–7, 9; 8:11).

Instar II (L2; Pls. 9–11; Fig. 4B). – Body length about 320 µm. Cephalic shield with distinctly defined margins (Pl. 9:3, 6–9; 10:2). Outline suboval, widest in the first third and narrowing towards the rounded posterior margin. Anterior and posterior margins slightly raised towards the middle (Pl. 9:9; 10:2). Shield moderately arched in profile, with the maximum height in the first third (Pl. 9:3). Measurements: length 230 µm, width 200 µm, height 80 µm. Area at greatest height slightly raised and with pores on its posterior margin (exact number unknown; Pl. 10:1). Shield extending behind the maxillulary segment, probably including the maxillary segment as well (Pl. 9:9).

Eye lobes similar to those of preceding stage, but slightly protruded by shield margin, about 35 µm wide and 25 µm long (Pl. 9:2, 3, 7–9; 10:2–5). Soft midventral area in all cases distorted, but knob or pore preserved (Pl. 10:4, 5). Labrum not well enough preserved for a detailed description (Pl. 9:1, 5, 7).

First antenna 120 µm long, with about 12–13 incomplete ringlets. Distal end similar to preceding stage (terminal segment in Pl. 10:7). Subterminal seta on anterior margin of distal cylindrical segment positioned at slightly less than the middle of the length of the segment (Pls. 9:8, 9; 10:6, 7).

Second antenna similar to instar I, but probably longer.

Median surface poorly documented, endopod distorted in all specimens at hand. Basipod with three ringlets on outer surface (Pl. 10:6). Exopod well-preserved in UB 912, having 16–17 rings but fewer setae (Pl. 9:8, 9). As in preceding stage, the setation does not conform with the segmentation of the ramus.

Mandible similar to adult one (Pl. 9:8, 9). Coxa well-defined, with two submarginal rows of denticles laterally (Pl. 10:8). Gnathobase prominent and slightly wider than in preceding stage, shovel-like, similar to adult mandible (Pl. 9:7, 8). Anterior margin of gnathobase straight, posterior margin gently curved towards the tip of the blade (Pl. 10:9). Posterior marginal spinules set off from the median group of approximately seven spinules, thicker. Gnathobasic seta large, with marginal setules (Pl. 10:10, 11).

First maxilla now inserting on anterior body portion, with gross shape already similar to adult limb, about 110 µm long. Held in a posteriorly flexed position in all specimens, flanking the trunk ventrolaterally. Protopod finely folded on outer surface, bearing four elongate endites medially that decrease in size progressively (Pls. 9:3, 5, 7–9; 11:1–6; Fig. 5B). All of them more produced posteriorly than anteriorly, and terminating in a denticulate spine (Pls. 9:5, 8; 11:4). Proximal endite bearing three marginal setae in addition to the posterior spine, seemingly belonging to shaft-like proximal portion. Subsequent protopodal segments distinctly separated also on outer surface, with one long seta anteriorly, fourth one carrying the exopod (Pls. 9:9; 11:3).

Proximal three endopodal segments wider than long, having pairs of setae medially. Terminal segment longer than wide and rounded apically, in contrast to later stages also with a pair of setae mediodistally and one thicker apical seta (Pl. 11:1-5; compare with distal ends of 1st antenna and endopods of 2nd antenna and mandible). Whole median surface of limb covered with numerous setules (in most cases broken off and indicated only as small dots). Exopod finely folded at its basis, one-segmented, reaching distally about to top of second endopodal segment. One thick terminal seta (Pls. 9:9; 11:3, 5).

Sternites of mandibular and maxillulary segments fused with one another, forming a post-labral sternal plate. Deep transverse furrow between larval head and hind body now behind maxillulary segment (Pl. 9:7, 8).

Trunk conical as in preceding stage, but slightly ventrally raised, about 140 µm long and 85 µm thick anteriorly. Buds of 2nd maxilla and first thoracopod feebly developed as small humps with slightly fringed distal rim (Pls. 9:7, 8; 11:1, 7). Anlage of 2nd maxilla slightly larger and more developed than that of first thoracopod, with denticles on distal surface. Both buds much less developed than rudimentary 1st maxilla of preceding stage.

Incipient furcal rami of about the same size as in preceding instar, with one acute spine and a smaller one laterally (Pls. 9:1–5; 7–9; 11:8). Length of dorsocaudal spine decreased to 25–30 µm. Posterior part of trunk and furcal rami furnished with numerous denticles, as in preceding stage. Few shallow transverse ridges on ventral surface of trunk between limb buds and rami may be indicative of incipient segments (Pl. 9:7).

Instar III (L3; Pl. 11:9, 10). – Stage represented only by three distorted specimens, which precludes a detailed description. Its recognition is based on the presence of three pairs of subtriangular rudimentary limbs, 2nd maxilla and two thoracopods, and on having a size between that of instars II and IV: cephalic shield, trunk, furcal rami, and total length are larger than in the preceding stage, whereas the dorsocaudal spine is shorter (Table 3).

Instar IV (L4; general views Pl. 12:1–5; details Pls. 12:6–8; 13; 14:1, 2; Fig. 4C). – Body length about 430 µm. Gross morphology with few significant changes. Length of cephalic shield three quarters of total body length. Shield now distinctly overhanging the ventral body anteriorly and laterally (Pl. 12:1, 2, 4). Due to this, eye no longer visible in profile. Eye lobes poorly preserved but apparently larger than in instar II (width to length 40/30 µm) and more bulging (Pl. 12:2). Posterior margin of shield slightly excavated in the middle (Pl. 12:1, 4), coalescing with the dorsal body wall (Pl. 14:2), similar to adult (compare with Pl. 3:8). Two furrows run dorsally from behind 1st maxilla and merge into margin of cephalic shield without fusing with one another. It is, however, not clear which segment this furrow is bordering.

Labrum prominent and ventrally projecting (Pl. 12:4–6), about 150 µm long. Shape more ovoid than in instar I and with 3–4 lateral rows of setules. Sternal surface furnished with numerous setules, as in instar I.

First antenna not preserved. Subsequent appendages larger than in instar II and similar to those of adult. Of the 2nd antenna only protopod and exopod are known in part (Pl. 12:1–4). Mandibular coxa thick and robust; gnathobase broader and more obliquely directed than in the preceding stages. Distal surface of gnathobase with numerous delicate setules, many of them arranged in rows parallel to the margin. Margin between prominent posterior and anteromedian spinules additionally fringed with many setules or denticles (Pls. 12:4–8; 13:1–3). Additional short spinule distal to large posterior one. Lower surface of gnathobase furnished with few denticles. Basipodal endite slightly flattened, similar to that of instar I, also with two rigid setulate spines and several setae around them (Pl. 13:1, 2). Shape of endopod most probably not changed, exopod composed of 10 segments with nine setae.

First maxilla 140–145 µm long, much as adult one (Fig. 5C). Shaft finely folded laterally and carrying the prominent, medially produced proximal endite (Pls. 12:4, 5; 13:4–8). Endite with six pectinate marginal setae and another seta above the row, flanked by a tiny spinule. Subsequent endite also slightly flattened, with two setae anteromedially and drawn out posteriorly into a horn-like protrusion which is tipped with a spine. A smaller spine positioned on anterior side of its basis. Third endite less bulging, with two anteromedian setae and a smaller one posteromedially. Fourth to seventh segments each with a pair of setae medially, the posterior seta being slightly thinner than the anterior one. Terminal endopodal segment with three setae apically in an outwardly running row. Exopod reaching to about the middle of the second endopodal segment, with two setae apically, surrounded by a number of setules and denticles (Pl. 13:4, 5, 8).

Trunk similar to preceding stages, in particular in dorsal and ventral views, but markedly larger (190 µm) and much higher than wide anteriorly (Pls. 12:1–5; 13:11; 14:1). In most cases preserved flexed ventrally against the anterior body portion at an angle of 120°–135° (Pl. 12:1, 2, 4, 5). Ventral surface gently bulging anteriorly and sloping towards the furcal rami.

Four pairs of limb buds developed on the anterior two thirds of trunk, representing 2nd maxilla and three thoracopods (Pl. 12:3–5). Anterior buds subtriangular, lobate, with more or less bifid distal margins. Median tips larger than outer ones, with one short apical spine, their sloping inner margins being slightly fringed, indicating the future development of endites and/or segmentation (Pl. 13:10, 11). A short spine on sloping outer margin indicates the future position of exopod, at least in the anterior buds. Buds progressively decreasing in size and state of incipient segmentation from front to rear (Pl. 13:9), last bud being only a small hump. All buds less developed than rudimentary 1st maxilla of instar I.

Furcal rami longer than in instar II (40–45 µm) and with three terminal setae: one seta medially and two thinner ones standing closer together laterally (Pl. 13:11; it may be that the preceding instar had already developed three setae too). Caudal end of trunk with anus as in all other stages (Pl. 14:1). Length of dorsocaudal spine unknown.

Instar V (L5; general views: Pl. 14:3–7; details: Pls. 14:8; 15; Fig. 4D). – Body of last stage of metanaupliar phase only slightly longer than that of preceding larva, about 450 µm; also other size changes fairly small (Table 3). However, stage can be easily distinguished from preceding ones by the presence of five pairs of trunk limb buds.

Cephalic shield in most cases preserved much wrinkled or distorted, but posterior margin excavated and posterolateral corners slightly more backwardly extended, already similar to presumed adult. Eye overhung by cephalic shield, lobes not significantly larger than in instar IV but probably slightly more separated from one another and farther posteriorly produced. Knob on narrow anterior end of midventral area present (Pl. 14:8). Labrum poorly documented, but groups of denticles on anterior surface and rows of setules laterally recognizable (Pl. 15:1, 2). Sternal surface with numerous setules (Pl. 15:2) and two swellings separated by a medial groove, indicating incipient paragnaths (Pl. 15:9).

First antenna similar to those of preceding stages, with about 12–13 ringlets and three distal segments. Second cylindrical segment with three setae, one at about two thirds of length of segment anteriorly and two setae distally, flanking the short terminal segment which carries the central two setae (Pl. 15:4).

Second antenna similar to that of adult. Outer surface up to exopod with distinct ringlets (Pl. 15:5). Coxa and basipod well separated from one another, both with long endites. Endites truncate distally, slightly more produced posteriorly than anteriorly and with two groups of setae or spines (Pl. 15:5). Proximal endopodal segment about as long as wide, produced into endite medioproximally, but much less than in the basipod. Second segment distinctly longer than wide. Penultimate segment again smaller, rounded distally, and with several setae of varying size, encircling the small terminal segment. The latter carries the two midmost setae (Pl.

15:5). Exopod with 17 annuli and fewer setae.

Mandible essentially as in adult and distinctly shorter than 2nd antenna (Table 3). Gnathobasic seta half as long as the gnathobase, similar in shape to that of preceding stages (Pl. 15:2, 6). Penultimate endopodal segment with row of five setae, terminal segment almost fused with the former segment (Pl. 15:6, 8). Exopod with 10 annuli and eight setae. 1st maxilla slightly longer than in instar IV (Pl. 14:4, 5; Fig. 5D; Table 3). Shaft pliable, only its endite being more sclerotized. All endites with the same number of setae as in preceding stage.

Trunk without significant changes in shape, with slightly concave dorsal line and gently convex ventral side, inclining posteriorly. However, surface finely wrinkled laterally between 1st maxilla and second pair of limb buds, similar to surface structure of the thorax of the presumed adult. Texture obscures the boundary between head and trunk (Pls. 15:7). Furcal rami again further elongated (Table 3; Pl. 14:5–7).

Five pairs of limb buds on trunk, extending backward to about three fourths of trunk length. Buds progressively decreasing in size and shape, showing some individual variation in development, particularly of the last pair (compare Pls. 14:5 and 15:7). Buds still bifid lobes, with no apparent change in shape compared to instar IV, except for a slightly larger size and slightly more pronounced tips (Pl. 15:7, 10; compare with 1st maxilla of instar I, Pls. 7; 8:10).

Life cycle

The postembryonic development of *Bredocaris* consists of five stages, subdivided in two steps.

The first step has a single instar with the three typical pairs of naupliar appendages and anlagen of the 1st maxilla on the hind body additionally. This indicates that it is not an 'orthonauplius' in the strict sense, but already a 'metanauplius'. As the individual length and development of the shield border varies considerably, (compare Pl. 7:3 and 6), it is possible that our first stage includes two instars, but without significant morphological differences. Such a habit

is not uncommon among arthropods. It occurs in chelicerate larvae (e.g. pantopod protonymphs [Behrens 1984]; larva D of Müller & Walossek 1986b) as well as in crustacean ones. In the parasitic poecilostomatoid copepods *Sarcotaces pacificus* and *Colobomatus pupa*, for example, all five naupliar instars are of the same size, and the first two stages show few differences in detail (Izawa 1973, 1975). The few data available, however, do not permit a further subdivision of the first stage of *Bredocaris*.

From the second instar, the maxillulary segment is incorporated in the anterior body portion, and the 1st maxilla is functional. During further development five more limbs appear successively on the larval trunk, but all remain at the level of rudimentary buds. Such delay of limb development is characteristic of the ontogeny of most Recent maxillopodan groups, in particular the Thecostraca and Copepoda. The second maxilla and first thoracopod appear simultaneously at one moult. Specimens with only the rudimentary 2nd maxilla have not been recognized, and it appears unlikely to us that such a stage ever existed. Again, the maxillary segment remains part of the larval trunk and does not become incorporated in the head, as the maxillulary segment.

The largest stage of *Bredocaris* can be clearly distinguished from the metanaupliar sequence by the development of a set of eight pairs of well-developed post-maxillulary appendages, including the 2nd maxilla and seven thoracopods. The most significant changes between the last metanauplius (L5) and the large stage are: enlargement of the shield, appearance of the head-trunk boundary behind the second maxilla, further elongation of the trunk and its tagmatization, addition of three more pairs of trunk limbs (acquisition of seven pairs of thoracopods), full development of all postmaxillulary limbs, and further size increase of more than 90%.

Many features remain remarkably constant throughout ontogeny, in particular the complete anterior head region, the 'naupliar head'. The 1st maxilla shows a stasis in shape throughout morphogenesis to the extent that it served as a

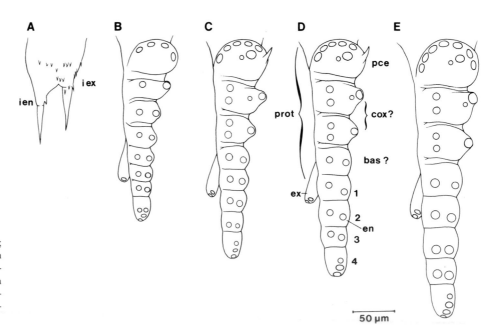

Fig. 5. Development of 1st maxilla; setae and spines omitted, except in (A). □ A. Rudimentary one of 1st instar. □ B–D. First maxilla of 2nd, 4th and 5th larval stage. □ E. Adult appendage (for discussion of the division of protopod, see pp. 12, 13).

useful character for the recognition of *Bredocaris* specimens within the collection, making it possible to combine larvae and largest stage (Fig. 5). Also other features link all developmental stages, such as the shape of the caudal end of the trunk with the rami and the characteristic pattern of denticles on appendages, labrum, and posterior trunk region.

There are few significant alterations of shape. In instar I only the sternite of the mandibular segment is developed. From the subsequent stage the sternites of mandibular and maxillulary segments fuse to a single plate. By the last metanaupliar stage, incipient paragnaths appear as a slight but distinctive pair of swellings on that part originally corresponding to the mandibular segment (Pl. 15:9). The paragnaths are raised further in the largest stage (Pl. 5:6). Although the shield seems to reach behind the maxillulary segment already in the first instar, it may be possible that this segment is added to the anterior body portion first with shifting of the 1st maxilla onto the anterior portion in instar II. The maxillary segment remains on the larval trunk until the largest stage.

The cephalic shield undergoes some change in shape during development. With progressive growth its posterior margin becomes more deeply excavated, while the posterolateral corners extend further backwards, loosely covering the anterior trunk segments. In the largest stage, the shield terminates behind the maxillary segment.

The progressively elongating trunk remains undivided throughout the metanaupliar phase. After moulting to the largest stage, the trunk is subdivided into a fleshy and only indistinctly segmented thorax and a smooth, uniform abdomen. In late metanauplii the anterior part of the trunk becomes more and more finely wrinkled, presaging the future surface texture of the thorax (compare Pls. 14:7; 15:7 and 1:3).

While most characters gain complexity or at least increase in size, the dorsocaudal spine is reduced progressively, a mode similar to certain Copepoda such as *Longipedia* (Gurney 1930; Onbé 1984). By contrast, in cirriped nauplii the spine may become increasingly more elongate (Barnes & Achituv 1981), a feature that is probably indicative of different functional needs of such an extension.

Stages between the last metanauplius and the 'trunk limb-stage' have not been found. It may be argued that intermediate stages had existed, simply not being recognized in the material. With few exceptions, however, the last nauplius/ metanauplius in the Maxillopoda, with still rudimentary post-maxillulary limbs, moults directly to the 'podid', with considerable changes of the morphology. Moreover, in the Thecostraca even the whole set of trunk limbs remains undefined externally, visible in late metanaupliar stage underneath the cuticle. With the moult to the cypris, all trunk appendages appear simultaneously (examples in Figs. 10F, 11). Consequently, the small difference between four and seven thoracopods in *Bredocaris* may have been easily bridged without intermediate stages.

The apparent similarity between the largest stage of *Bredocaris* and post-metanaupliar stages, the 'podids', of Recent Maxillopoda may be evidence to assume that the former represents an immature stage rather than the adult. According to Boxshall (personal communication), in terms of trunk differentiation the largest stage of *Bredocaris* is equivalent to about the third copepodid stage of Copepoda, which has seven trunk segments and an unsegmented pleotelson. However, *Bredocaris* has one more pair of trunk appendages than all 'podids' *and* than the adults of other Maxillopoda, which have at most only six pairs of thoracopods. Even if our largest stage is not the adult, this argument would still be valid, and would imply that seven pairs of thoracopods is the final state. Again, there are no indications of structures of larval shape in the largest stage. This is particularly true of the trunk limbs, where one would expect a reduction in size and armature in the posterior ones. The lack of distinct thoracic tergites and only faint segmentation of trunk and limbs is considered as secondary and in accord with attainment of a highly flexible trunk as a special life strategy.

The major arguments for the recognition of the largest stage of *Bredocaris* as the adult are in our view:

- the body segmentation: addition of the maxillary segment to the head and tagmatization of the trunk into thorax and abdomen in this stage,

- and the full development of the 2nd maxillae and seven pairs of thoracopods, and their equipment with enditic lobes and spines on the median limb surface.

On the basis of the material at hand, the larval sequence discovered is concluded to be complete. It comprises five metanaupliar stages, but lacks the 'podid' phase. Such condensation of ontogeny, with complete suppression of the subimaginal phase, is not a crucial difference to the ontogeny of Maxillopoda, in particular the Thecostraca (see p. 28).

Summed up, development of *Bredocaris* is anameric in the sense of Kaestner (1967, p. 921), but shows several significant alterations from the gradual type:

- it starts with an early metanauplius, having rudimentary 1st maxillae;

- the 1st maxilla is functional in the 2nd metanauplius;

- 2nd maxilla and 1st thoracopod appear simultaneously at this stage;

- three more thoracopods are added gradually, but all five post-maxillulary limbs remain rudimentary (delay);

- if the developmental sequence recognized is accepted to be complete, the fifth metanauplius moults directly to the adult and the whole set of functional trunk limbs – 2nd maxilla plus seven thoracopods – appears after one moult.

Aspects of functional morphology and mode of life

Size and articulations

Bredocaris is in the same size range as various tiny Recent benthic or epibenthic crustaceans, in particular those of the meiobenthos of flocculent bottom layers or the mesopsammon. Elements of these associations, even if they belong to different higher taxa, share a small size, worm-like shape and high flexibility, and they reveal many similarities in basic life habits, such as locomotion, alimentation, orientation, reproduction and life cycle (cf. Noodt 1974; Siewing 1985, his Fig. 937). Progenetic development and neoteny are also common phenomena (Noodt 1974, Schminke 1981). Crustaceans are abundant elements of these associations, such as Bathynellacea among the Malacostraca (Schminke 1981) and members of all major non-malacostracan groups.

In accord with small size, a firmly sclerotized exoskeleton is not necessarily developed, and turgor pressure gives rigidity to the exoskeleton. Joints are often only feebly defined. Whereas, for example, the long slender trunk of the *orsten* Skaracarida is distinctly segmented into subconical annuli which are separated by arthrodial membranes (Müller & Walossek 1985), the thorax of *Bredocaris* lacks distinct tergites and segmentation is only faint. The cuticle is finely wrinkled in all specimens, also in inflated ones, indicating its softness. It is, however, likely that slight shrinkage after death caused some enhancement of the texture.

The softness of the thorax may have permitted a similar high degree of flexibility of the trunk into various directions, shown in Fig. 6, which is reconstructed from actual specimens, combined with illustrations of curvature and limb position of the very similar cypris Y larvae (Bresciani 1965; Schram 1970a; unpublished SEM-micrographs kindly submitted by Dr. Ito, Kyoto).

Dorsal flexure is assumed to have been less than the ventral one. This assumption is mainly derived from the mode of the insertion of the appendages (Fig. 6A). In the maximum ventral position of the trunk, the thoracopods were put closely together and held parallel to the abdomen (Fig. 6C; see also Pl. 2:1–5 and Schram [1970a], his Fig. 1, for a Y larva). The lack of a distinct joint between the slightly firmer abdomen and the thorax points against separate movements of the former.

All appendages are only flexibly articulated to the body, with exception of the mandible. The 1st antenna, about as large as the labrum, inserts more medially than the subsequent limbs (Figs. 2, 12, 14). Its preservation in different positions, from being anteriorly stretched to a position between labrum and endopod of 2nd antenna, suggests a major to and fro swing, with slight outward–inward direction (see also Figs. 12, 14, 15, 17). Its flexibility is enhanced by the lack of segmentation on the posterior side of the proximal three quarters of the appendage (Fig. 3A).

The 2nd antenna (Fig. 3B) inserts lateral to the labrum (Figs. 2, 4E). Its protopod is slightly posteromedially oriented, which may have permitted not only an anteroposteriorly directed swing of the limb, but also some inward movement to pass the enditic spines and setae alongside the labrum medially and orally during the back swing of the limb (Pls. 7:1, 8:7; 12:4, 5; Figs. 2, 13, 17). The outer surface of the protopod is annulated as in the 1st antenna, suggesting good flexibility (Pls. 3:3, 4; 4:1, 2). The exopod arises from the sloping distal surface of the basipod, with its setae pointing mediodistally, as is typical among Crustacea. The well-developed segmentation of the ramus suggests the ability to spread the setae in a fan, the direction of which is predicted by their posterodistally directed sockets.

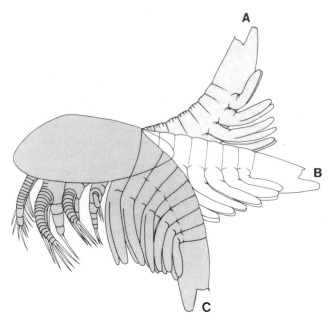

Fig. 6. Scheme of adult *Bredocaris* in different kinds of body flexure. □A. Upward curvature of the trunk. □B. Trunk stretched. □C. Supposed maximum ventral flexure of trunk.

In several specimens the antennal and mandibular exopods are preserved curved posteriorly (e.g. Pl. 4:2). Flexure may have been less pronounced during life, however. Perryman (1961) demonstrated how much shrinkage as well as inflation during fixation can influence the general shape and the position of appendages (her Figs. 18, 19, 27, 28).

The mandible (Fig. 3C) inserts immediately behind the labrum (Figs. 2, 4E). The large, abaxially oriented joint between the firmly sclerotized coxa and the body is well-

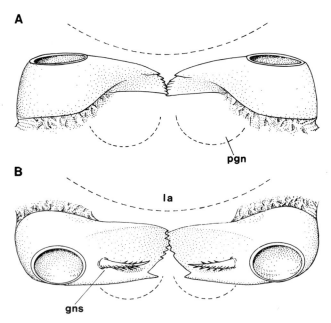

Fig. 7. Movements of the mandibular coxae in accord with limb movements; positions of labrum and paragnaths indicated by stippled lines. □A. Coxae anteriorly directed, gnathobase almost vertically oriented against posterior edge of labrum. □B. Coxae posteriorly directed at the end of the back swing of the mandibles, gnathobases approaching the gnathobases.

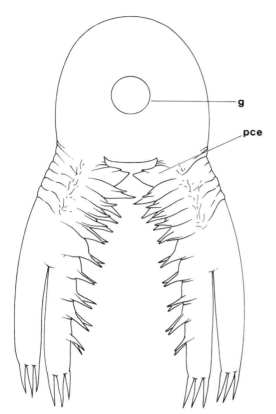

Fig. 8. Cross-section of anterior trunk region with ventrally stretched pair of thoracopods.

defined. This may have not only permitted a wide anterior–posterior swing of the limb but also considerable outward–inward grinding movements (Pls. 4:4; 10:9). Another joint between coxa and basipod (Pls. 4:5; 10:8; 13:1–3) permitted grinding and stuffing activities of the coxal gnathobase separate from the locomotion and food collection of the distal limb portion. The similar construction of the distal portions of 2nd antenna and mandible indicates that the 2nd antenna did not contribute alone to swimming and food collection, but both limbs acted together.

The 1st maxilla (Fig. 3D) inserts almost abaxially, but slightly more inward to the mandible (Pl. 5:6; Fig. 2). In larval specimens the protopod may have been slightly more obliquely anteriorly oriented than in the adult (Pl. 12:4), with the enditic setae pointing anteromedially. In all specimens at hand this limb is posteriorly curved, mostly lying against the trunk in the larvae or on the trunk limbs in the adult (Pls. 1:3; 4:7–9; 9:7–9; 11:1–5; 14:2; Fig. 4). The finely folded outer edge of the shaft suggests good flexibility of the limb and wide range of action, in particular an outward–inward swing. It is not very likely that its large proximal endite was capable of separate movements.

The 2nd maxilla and the anterior thoracopods (Fig. 3E–I) are almost ventrally or slightly outwardly directed (Pls. 1:1–4; 2:1–7, 9; Figs. 2, 6, 8). Towards the posterior, they become progressively more laterally positioned (Pls. 1:3; 2:4, 5, 9). Although the body tapers progressively, due to this shifting of the insertions onto the sides of the body, the median path does at least not narrow posteriorly (Figs. 2, 14, 15, 17A). The change of positions may also have influenced the range of action of the limbs within the series.

Insertion and orientation is very similar to the thoracopods of cyprids of Facetotecta (Schram 1970a; Fig. 10I), Ascothoracida (e.g. Brattström 1948; Grygier 1984b), and Tantulocarida (Boxshall & Lincoln 1983, 1987). The limbs of these forms are slightly better articulated and segmented. The low degree of external segmentation of the trunk limbs of *Bredocaris*, remarkably similar to trunk limbs of Branchiura, having long fleshy bases (Tokioka 1936, Fig. 12A; Shimura 1981; 'accordion-like plates' according to Grygier 1984b), suggests slightly different functional strategies in the use of these limbs. The major axis of swing may have been forward–backward (see also Figs. 2, 6), the flexible shafts, however, may also have permitted slight outward–inward movements.

Locomotion and feeding

Interpretations of motility and feeding habit are derived from the different preservation of the specimens, from comparisons with similarly constructed extant crustaceans, and the clay model. The similarities between *Bredocaris* and various members of the Maxillopoda, in particular their larval stages, serves as additional source for comparisons.

Living at low Reynolds number profoundly affects the behavior pattern of small-sized animals. Using high-speed cinemicrography, recent studies on crustaceans demonstrated how classical models of behavior may differ from the new evidence, since interpretations were often much influenced by observations on larger forms operating at higher Reynolds numbers (cf. Koehl & Strickler 1981 for Copepoda). In particular the small larvae are surrounded by a thick viscous boundary layer, while elongation of the body during development enables the animal to achieve a better hydrodynamic shape and to reduce the boundary effects (Barlow & Sleigh 1980 for *Artemia*). Interpretations and comparisons merely on the basis of external morphology must remain limited, since different forms may have developed different strategies. It is difficult to access in full detail the true range of action of the appendages and the correlation between their beats, whether it was synchronous or out of phase. Thus, the model gives only a more generalized idea of possibilities of the locomotory and feeding apparatus of *Bredocaris*.

Larvae. – Gauld (1959) and Sanders (1963) presented detailed comparisons of the various Recent crustacean larvae. At least two clearly different types of nauplii were differentiated: the branchiopod type and the non-branchiopod type. While the mechanism of swimming is rather uniform, there is a strong difference in particular in the use of the appendages. Branchiopoda use exclusively their 2nd antenna for locomotion, whereas all other types in general use all three naupliar limbs.

Various early crustacean instars are non-feeding and thus have reduced feeding apparatus on one or all three appendages. This is particularly true of the free living malacostracan nauplii. Similarities in design are thus apparent only in very basic features. In the nauplius of *Artemia*, the 2nd antenna is not only the major locomotory organ but also the principal food collector. The typical features of this branchiopod type, such as masticatory spines on the antennal protopod, reduced 1st antenna, hinged setae, mandibles concerned only with feeding, and a very mobile, elongate and posteriorly directed labrum, are already known from the larvae of the Devonian lipostracan *Lepidocaris* (Scourfield 1926, 1940).

Compared to this type, the larvae of *Bredocaris* clearly belong to the non-branchiopod nauplius type, which is mainly characterized by a longer and segmented 1st antenna, feeding structures on the 2nd antenna, and a biramous mandible (Sanders 1963). In this type the 1st antenna and the mandibular rami assist in food gathering. In accord with this, the 1st antenna of *Bredocaris* is about as large as the labrum, multi-articulated and with distal and medial setation. When it moved backwards, the setae could reach the postoral chamber (see also Fig. 17). Morphology and movability indicate its assistance in both swimming and trophic function. Already the initial instar was a feeding stage, recognizable in the armatures of 2nd antenna and mandible, and the development of mouth and anus.

The 2nd antenna is the longest appendage, indicating its importance in locomotion and food gathering. Both, coxa and basipod have long enditic processes with terminal spines. During the backstroke these move alongside the labrum and finally come to lie on the mandibular gnathobase. While moving posteriorly they pass the rows of setules at the sides of the labrum (Pls. 8:3; 12:6). The endopod is also prominent and furnished with setae of varying size. The larger of these may have assisted in food collection, similar as described for other larval Crustacea (e.g. Gauld 1959; Barlow & Sleigh 1980).

The rami of the mandible are shorter than those of the 2nd antenna but well-developed. The basipod is prominent and armed with rigid masticatory spines. The coxal gnathobase is conspicuous from the first larval stage and better developed than in all Recent maxillopodan larvae (examples in Fig. 11).

From the second stage, four pairs of appendages were involved in the feeding process, including the first maxillae. They surround the postoral feeding chamber, which was bordered by the labrum anteriorly and the slightly ventrally flexed trunk posteriorly (Pls. 7:3; 9:3; 12:4, 5; 14:7; Fig. 4B–D). The ventral curvature of the larval trunk has also been described for the copepod feeding apparatus (Marcotte 1977), confirming that this habit was an original life position. The 1st maxilla does not change significantly throughout further development (Fig. 5). The rudimentary habit of the exopod and well-developed median setation of the whole limb indicate that this limb had an almost exclusively trophic function, but the setules on the posterior edges of the distal endopodal segments and the exopod may also indicate some grooming function of this limb.

Adults. – The presumed adult of *Bredocaris* obviously retains the complete metanaupliar locomotory and feeding apparatus with no significant alterations (Figs. 2, 4E). Thus all assumptions on the larval feeding process may also be attributed to the adult.

Motion of 2nd antenna and mandible may have been similar to the process described for the larval 2nd antenna of *Artemia* by Barlow & Sleigh (1980). Disregarding possible phase differences between the appendages, the beat cycle may have started with the appendages directed anteriorly.

The ramal setae spread into fans. During the propulsive or power stroke, the rami were not only moving backwards over a wide angle, but may also have curved slightly medially, finally rotating towards the mouth (Fig. 17B). Most likely, due to space competition between the two limbs, their rami may have curved inwardly as much as the 2nd antenna of *Artemia*. During the recovery stroke in particular the exopod and its setae were flexed posteriorly to reduce the drag. According to Gauld (1959) the setal fan also acts as a sweep net to capture food particles. Barlow & Sleigh (1980) pointed out that this is true only of the larger, feeding larvae of *Artemia*, while Fryer (1983) stressed the purely natatory function of the exopodal setae.

Bredocaris may have collected food particles of moderate size, which were swept behind the head by the rami of 2nd antenna and mandible, supported by the 1st antenna. There the particles may have been grasped by the 1st maxilla and brought into the post-labral chamber, where they were shovelled anteriorly between the mandibular gnathobases mainly by the flattened proximal endites of the 1st maxilla and their marginal pectinate setae. The gnathobases finally stuffed the particles into the mouth (Fig. 17B).

When the mandible swung anteriorly, it tilted progressively in an almost vertical position, approaching the posterior side of the labrum with their distal surfaces (Fig. 7A; see also Pls. 9:1; 12:4–8; 13:1, 2). A slight overlap of the gnathobases and marginal spinules permitted them to crunch particles and to stuff the food into the atrium. Again, brushing off the food particles from the grinding plates may have been implemented by the setules on the distal surface of the plates (Pls. 12:7; 13:3) and the gnathobasic seta (Pls. 10:10, 11; 15:2). Similar setae are developed in all other crustacean groups as well, serving to sweep particles off the distal edge (some examples in Sanders 1963; see also Fig. 11). In the posterior position of the mandible the angled gnathobases came to lie on the paragnaths (Pls. 9:7, 8; 10:9; Figs. 7B, 17B). The paragnaths, over which the gnathobases were moved, are not very prominent, and the groove between them is accordingly shallow.

In addition, the adult possesses a series of functional trunk limbs, including the 2nd maxilla and seven pairs of thoracopods. Uniform trunk limb series are developed in various Crustacea and their larvae, but differ considerably in detail. In *Lepidocaris*, for example, the trunk limbs are basically similar to one another, but the series is divided into groups with different function: the anterior three pairs are foliate and phyllopod-like, functioning mainly for filtration and food scraping, while the posterior set is less furnished with medial setation and, with its uniform paddle-shaped rami, may have served mainly for locomotion (Scourfield 1926, Pl. 23; Fig. 11U herein; see also Fryer 1985). All trunk limbs, except the last three, have a distinct proximal endite, as in *Bredocaris*. However, in the latter form these are simple lobate outgrowths with few short spines only. Again, although the trunk limbs of *Bredocaris* have numerous endites, in gross shape they resemble more the locomotory posterior limbs of the trunk limb set of *Lepidocaris*.

The bottom-crawling Cephalocarida are not filtering, as variously suggested (Lauterbach 1980, 1983, 1986), but are non-selective deposit-feeders (Sanders 1963, p. 9). Setation of the cephalocaridan trunk limb (Fig. 11T) is not specialized for filtration (e.g. pectinate filter setae), but considerably better developed than in *Bredocaris*. Their endopod is distinctly divided into five or six podomeres, the distal one forming a claw. With these claws food is scratched from the bottom and passed upwards into the food groove between the limbs, where it is brushed anteriorly towards the specialized first maxilla by means of metachronal beating. The exopod is two-segmented, and there is a further exite called 'pre-epipod' or 'epipod' (in respect of its insertion close to the exopod or even at its basal segment, however, neither of the terms applies in the strict sense).

This construction clearly contrasts with the trunk limbs of *Bredocaris*. Here the median spines, projecting into the narrow path between the trunk limbs, form a regular structure (Figs. 2, 3E–I, 15, 17), but they are rather short and naked. Furthermore, special aids for grasping, such as movable limb parts, long setae, scratching, such as rigid denticulate spines, filtration and/or food transport, such as setulate setae, setulate endites and a median food groove, are all absent. Particularly in respect of the short proximal endites, lacking transporting setae as on the 1st maxilla, there is little evidence for a production of food currents anteriorly along the ventral body surface, as would have been necessary for filtering. Neither is there any evidence for a simpler mechanism similar to that of the Cephalocarida. (Regarding filtration, cf. Fryer 1983, 1985.)

It cannot be excluded that the trunk limbs of *Bredocaris* may have served as a coarse cage to trap particles which were then transported anteriorly by means of their median armament, but such a structure could not have been very efficient. Appendage morphology, however, shows that *Bredocaris* was clearly not filter-feeding.

The trunk limbs of *Bredocaris* are slender and flattened, being composed of prominent protopods and two paddle-shaped rami about equal in size. Similar thoracopod series are present in various small crustaceans, and in particular in the cyprids of Ascothoracida (e.g. Brattström 1948, Figs. 7C–E, 13A, B, 15D, 19F; Grygier & Fratt 1984, Fig. 5F), Facetotecta (Schram 1970a, Fig. 4C; Ito 1985, 1986b, Fig. 2E), Tantulocarida (Boxshall & Lincoln 1983, 1987), and Branchiura (Shimura 1981). As in all these forms the trunk limbs are mainly concerned with swimming, in our view, the shape of the trunk limbs of *Bredocaris* fits better with a major role of the trunk limbs in locomotion. This would be also in accord with the necessity to stabilize the long adult trunk.

Possible movements were studied with the plasticine model. The mode of insertion and orientation of the trunk limbs suggests metachronal beating of these, as is common in Crustacea with a long series of swimming legs (e.g. Barlow & Sleigh 1980; Hessler 1985). As phase differences are impossible to formulate, Fig. 16A–C presents a simplified scheme of the possible beat cycle of the trunk limb series.

According to Hessler (1985), the protopods ('peduncles') commonly serve as major power-generating elements, while the rami followed them more or less passively. As peduncle and rami articulate with one another, drag can be reduced on recovery by flexure of the distal limb portion (cf. Hessler 1985, Fig. 7). The construction of the trunk limbs of *Bredocaris* obviously permitted such mechanism, although being almost devoid of segmentation and having a flexible shaft. The slightly better sclerotized basipod may have represented

an intermediate element to find a concession between high flexibility of the limb and rigidity to withstand the viscosity of the surrounding medium during the beat. Moreover, it also enabled flexure of the distal portion posteriorly during the recovery stroke at its proximal border (Pl. 16; Figs. 4E–I, 16B).

In various Crustacea, the efficiency of locomotory limbs is enhanced by development of marginal setation, in particular at the distal end of the rami. Efficiency can also be enhanced by the large size of the limbs relative to the body and prominence of the protopods. As the trunk limbs of *Bredocaris* lack setation, the latter strategy may have been followed. Again, several crustaceans have developed special structures in order to facilitate a joint beating of the trunk limbs. Copepoda and Facetotecta, for example, bear special coupling devices between the members of a limb pair, while in Tantulocarida special coupling spines on the thoracopodal endites serve the same purpose (Boxshall & Lincoln 1987). In the praniza larvae of gnathiid isopods, peracaridean malacostracans, spines on the inner margins of the protopods serve as coupling aids to connect the pleopods (Cals 1978, Fig. 4). It is possible that in *Bredocaris* the median spines of the trunk limbs, or at least the more proximal ones, may have functioned similarly, i.e. to couple the limbs during motion, if necessary. Easy disconnection, on the other hand, could have retained the separate movability of the limbs of either side for steering.

The abdomen is slightly firmer than the thorax. It may have stabilized the body from bending, while the furcal rami and their setae served as rudders.

In conclusion, adult *Bredocaris* had two functional systems operating in conjunction: the metanaupliar set of cephalic appendages and the trunk limbs. Both sets join at the 1st maxillae, which were not involved in locomotion. It is likely that the feeding apparatus of the metanaupliar cephalic appendages predominated that of the trunk limb series by far.

The mode of motion and feeding of *Bredocaris* is apparently different from that of other *orsten* forms, such as the Skaracarida, which are cephalo-maxillipedal feeders, or *Martinssonia elongata*, a bottom-dweller that stirred up food with its limbs and pleotelson-like tail. The phosphatocopine ostracodes may have been basically suspension feeders, while *Dala peilertae* and *Rehbachiella kinnekullensis* (both Müller 1983) probably were true filter feeders.

Life habit and habitat

The retention of many larval features in the largest stage indicates similarities in the mode of feeding and the preference in the source of food. A more or less permanent swimming mode of life appears more likely to us than other modes of locomotion, such as crawling, as in the Mystacocarida among the Maxillopoda (Lombardi & Ruppert 1982) or the Cephalocarida (Sanders 1963). Swimming must have been different between larvae and adult due to elongation of the trunk in the adult and development of trunk limbs, but other life habits and the habitat may have been shared. This assumption may be corroborated by the common occurrence of young and adults in the samples (Table 1).

The habitat of *Bredocaris* may have been a flocculent bottom layer, with low currents but high nutrient content. It is assumed that the majority of *orsten* arthropods lived benthic or epibenthic in such environment that provided many different ecological niches (Müller & Walossek 1986c). With regards to the possible swimming mode of life, *Bredocaris* may not have lived within the uppermost zone of such a flocculent layer but swam around close above it.

Affinites

Remarks on the Maxillopoda

In the last few years important new evidence concerning this crustacean subclass, established by Dahl in 1956, has been brought up, in particular for the Ascothoracida (Grygier 1983b, 1984b, 1987a for further references), the tiny parasitic Tantulocarida (Boxshall & Lincoln 1983, 1987), and the so-called Hansen nauplius and cypris Y larvae (Ito 1985, 1986a–c). The latter larvae are now classified as Facetotecta (Grygier 1985a). Valuable attempts have been made to portray the general plan of Copepoda (Boxshall, Ferrari & Tiemann 1984), Ascothoracida (Grygier 1983a), the Maxillopoda (Newman 1983), and to define synapomorphies for the subclass (Newman 1982; Grygier 1983a, 1984b, 1987a; Boxshall & Lincoln 1987). The value of naupliar morphology for ingroup analysis has been demonstrated by Grygier (1987a). Recently the extinct order Skaracarida has been added to the subclass by Müller & Walossek (1985), providing a hitherto unknown bauplan within this group.

With this new information, the current discussion of the validity of the Maxillopoda as a taxon has focused on the recognition of at least a 'core group' around the shield-bearing Thecostraca. This name, first introduced by Gruvel in 1905, was revitalized by Grygier (1984b, formally in 1985a) to include the Ascothoracida, the Cirripedia, and the Facetotecta. According to Boxshall (personal communication) this core group includes the Ascothoracida, the Cirripedia, the Facetotecta, and moreover the Tantulocarida, probably the sister group of the Recent members of the Thecostraca in the sense of Grygier (1985a), and the Branchiura (Boxshall & Lincoln 1987).

Principal candidates for synapomorphies of this core may be the possession of basically seven thoracomeres, each originally bearing a pair of limbs, and the presence of (at least in the male) gonopores on the seventh trunk segment (in branchiurans the abdomen is reduced). According to Grygier (1987a) a possible synapomorphy of the Maxillopoda may be the basically eight-segmented 1st antenna (but see below) while thecostracans are characterized by the common possession of attachment devices on the first antenna and lack of postmaxillulary naupliar limb buds (but see below).

The relationships within the Maxillopoda, even between the members of the core group, are still not fully resolved. Two characteristic features of the thecostracan core group, the existence of a modified pair of appendages and, at least, the male gonopores on the seventh trunk segment, link this group with the Copepoda (Boxshall, personal communication).

The interrelationships of the Branchiura, the Mystacocarida, the fossil Skaracarida, and the Ostracoda remain problematical. Although evidence for inclusion of the latter group into the Maxillopoda has been variously presented (Elofsson

1966; Schulz 1976; Andersson 1977; Eberhard 1981), relationships to the other orders are unknown. Branchiura are highly modified due to ectoparasitism and possess a unique ontogeny. Mystacocarida (Dahl 1952; Hessler & Sanders 1966; Hessler 1971) show a remarkable mixture of primitive and derived features, and, according to Hessler (1982) and Newman (1983), adult Mystacocarida appear very 'juvenilized'.

The strange, endoparasitic Pentastomida have also been assigned to the Maxillopoda (cf. Grygier 1983a, 1984a for further references), but they are not considered herein.

Beklemishev (1952, 1969; fide Hessler 1982 and Schram 1982), included Copepoda, Cirripedia, Branchiura, and Ascothoracida within the taxon 'Copepodoida', which refers to the Copepoda as the central group. None of the names Maxillopoda and Copepodoida is fully satisfactory, since they refer to characters only of certain members: not all groups are copepod-like, and not all have maxillipeds. The widely used name Maxillopoda is preferred here, until a more appropriate name is proposed.

Comparisons with models of the 'urmaxillopod'

Various attempts have been made to reconstruct the ancestral maxillopod. Suitable for comparisons with the new order are the models of Grygier (1983a, Fig. 2B) and Newman (1983, Fig. 2G), which are not based on 'urcrustacean' morphology. The reconstruction of Tiemann (1984, Fig. 2) refers more to the urcopepodan plan rather than to the urmaxillopodan plan and is not compared here.

Although derived from different points of view, the models of Grygier and of Newman show remarkable similarities with the Orstenocarida. Grygier's model is based on a generalized, non-parasitic ascothoracid, as in his view 'urascothoracids' may not have been far from the base of maxillopodan evolution. Newman reconstructed his urmaxillopod as a certain instar of the larval sequence of urmalacostracans, which he termed the 'premancoid caripodid'.

Similarities between these models and Orstenocarida are most obvious in basic body segmentation, five limb-bearing head segments, covered loosely by a large shield, a thoracic region with seven paddle-shaped appendages, and probable presence of a compound eye. Discrepancies between the models and Orstenocarida, such as the size of the cephalic shield, some details of head and appendage morphology, and the caudal end of the trunk, are at least in part due to the restriction of the models to particular Recent groups. In all, both models give support to the assumptions of the inclusion of Orstenocarida within the Maxillopoda.

According to Hessler & Newman (1975) a large number of segments and distinct segmentation are likely to represent the more primordial condition. With regards to the distinctly segmented trunk of both models, the Orstenocarida seem relatively derived. This and the mode of development indicate that Orstenocarida can provide useful information of the urmaxillopod morphology, but may not represent the ancestral stock from which the Recent members of this subclass were derived. It is possible that they derived by further progenesis from better segmented ancestors, probably from a form which was segmented as the 'urascothoracid' of Grygier.

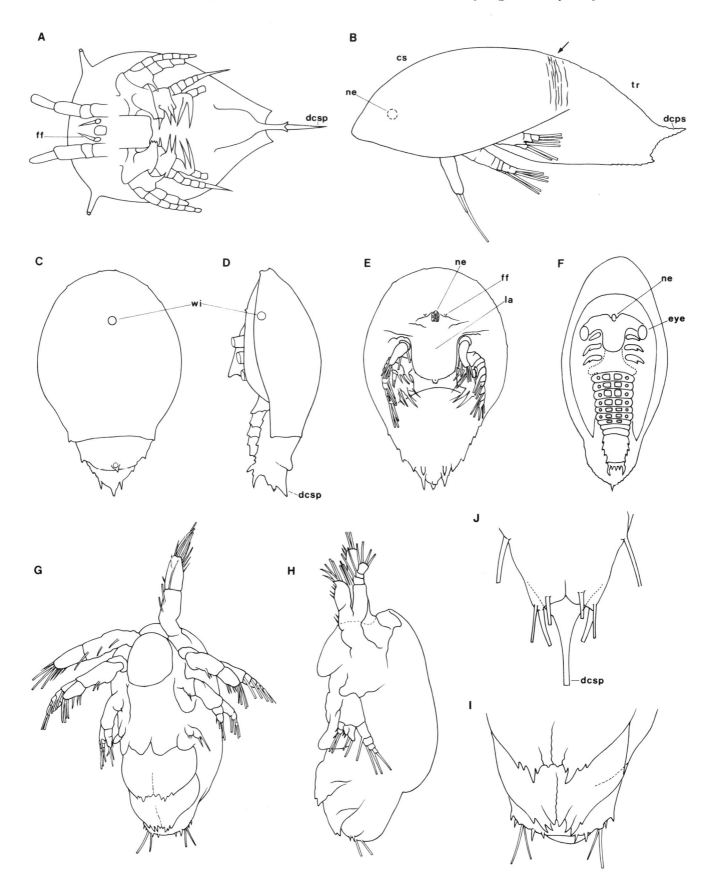

Fig. 9. Selected naupliar and metanaupliar larvae of Maxillopoda; most of setae shortened. Sizes not scaled. □ A. Nauplius of the cirriped barnacle *Balanus balanoides* (after Walley 1969, Fig. 1; internal structures omitted). □ B–E. Nauplii of Recent Facetotecta (exoskeletal sculpture omitted): B nauplius type VII; arrow points to boundary between shield and trunk (after Ito 1986a, Fig. 3B); C–E facetotectan nauplius type I in dorsal (C), lateral(D), and ventral (E) view (after Ito 1986a, Fig. 1A–C). □ F. Facetotectan metanauplius II still within the shell of stage II, ventral view (after McMurrich 1917, Fig. 10). □ G–I. Metanaupliar stage IV of the poecilostome cyclopoid copepod *Neanthessius renicolis* (after Izawa 1986, Fig. 23A–C): G lateral view, H subventral view, I detail of hind body with rudimentary thoracic limb buds □ J. Caudal end of 3rd larval stage of the harpacticoid copepod *Longipedia americana* (after Onbé 1984, Fig. 8).

ly specialized due to parasitism or a sessile mode of life, the position of the Orstenocarida could be at best only rather close to its base or adjacent to it, at a level when the group was not yet parasitic. This assumption is supported by the even greater similarity between Orstenocarida and the generalized plan of Ascothoracida, the most primordial group among the Thecostraca (Grygier 1983a, 1984b, 1987a) and in particular the immature stages of all members, which have apparently retained many conservative features. Examples of such instars are given in Figs. 9 and 10 (see also Grygier 1987b).

Facetotectan nauplii (Fig. 9B–E) are not only of the same size and shape, but also the short, simple shield has the same outline as in larval Orstenocarida and is also demarcated from the hind body by fine folds dorsally (Fig. 9B). Furthermore, a deep suture separates the naupliar head from the conically tapering hind body, unlike the copepod nauplius (Fig. 9G–I; see also Schram 1970b, 1972). As the facetotectan metanauplii (Fig. 9F), late ascothoracid metanauplii show anlagen of trunk limbs underneath the trunk cuticle (Fig. 10G, H). These buds are not only oriented similar to the limbs of adult Orstenocarida, they also have three to four distal setae on their paddle-shaped rami (cf. Brattström 1948, Fig. 13B, 15D, 19F; Wagin 1954, Fig. 7; Grygier & Fratt 1984, Fig. 5F). At this stage the 2nd maxilla is still leg-like (see also Wagin 1946).

Another indicator of close affinities to the thecostracan core group is the larval development, with tendency to reduce the larval as well as the post-naupliar phase. Copepoda may show condensation of the larval series, particularly in parasitic forms (e.g. Izawa 1973, 1975, 1986), but basically they possess the largest series of instars among the Maxillopoda, with six larvae and five 'podids'. Of the Thecostraca, the Ascothoracida have reduced the 'podid' phase to one or two cypris stages, in the Cirripedia is one left. Development of the Facetotecta is not completely known, and Tantulocarida hatch directly as 'podids', termed tantulus larva. This instar undergoes no further moults (Boxshall & Lincoln 1987). The presumed complete suppression of the 'podid' phase but maturation of a juvenile in Orstenocarida may thus be interpreted as progression on this way. Development of Branchiura is highly modified: generally the hatched larva is already similar to the adult and infective, and the elongation of the juvenile phase (eight stages; Shimura 1981) is obviously an autapomorphy of the group.

In contrast to Orstenocarida, the free thorax of thecostracan 'podids' is well-segmented, often bearing tergites. Interestingly, in the facetotectan *Hansenocaris*, for example, the three free anterior thoracomeres lack such pleurae (Ito 1986b), probably indicative of a tendency to reduce segmentation.

The shield of Facetotecta has been described as cephalothoracic, since the first thoracomere is fused to the head (Ito 1985, 1986b). However, Ito (1986b) remarked that in *Hansenocaris tentaculata* 'the fused first thoracic segment (with leg1) ... [is] ... demarcated from ... [the] ... cephalon by a ventral suture'. In our view the dorsal border requires further examinations, as it is unclear how far back the shield is fused with the body dorsally (cf. description of this feature for *Bredocaris* on p. 8). The stippled line in Schram's illustration (1970a, Fig. 1) indicates that the first thoraco-

mere is demarcated from the head. In this, the presence of a cephalothoracic shield of Facetotecta is not unequivocal.

Affinities of the Orstenocarida with Mystacocarida, Copepoda, and Skaracarida are apparently less close, although there are similarities with all of these groups, even in minor details. This uncertainty is not lastly due to their unresolved relationship to the thecostracan core group and affects not only the position of *Bredocaris* but also makes it difficult to interpret the polarity state of various diagnostic features, for example the shield, the rostrum, the two eye types, and the fate of the seventh thoracomere.

Copepoda reveal some affinities to the thecostracan core group in the shared position of the male gonopores. However, the mode of specialization of the seventh pair of limbs is different between the two groups (Boxshall & Lincoln 1987), and the position of the gonopores may be apomorphic of Maxillopoda but symplesiomorphic for its subgroups. If the Copepoda, with the largest series of instars (a clearly plesiomorphic feature with respect to the thecostracan core and *Bredocaris*), are included within the core group, this would imply only that all are maxillopodans. The exact interrelationships, however, remain unresolved.

Consequences, with notes on DALA PEILERTAE

The new order provides further evidence for the Maxillopoda as a valid taxon. It not only confirms various of the suggested characters diagnostic of this subclass, but also adds new information on details. Its morphology may also provide support of neoteny as a major evolutionary force for the derivation of the Maxillopoda from their multi-segmented ancestors.

The review confirms the position of the Skaracarida within the Maxillopoda, as has been suggested previously (Müller & Walossek 1985). Skaracarida show a mosaic of features distributed throughout the maxillopodan groups. However, if the presence of the tergite-bearing one or two anterior thoracomeres is a prerequisite for the formation of a cephalothorax, this feature, together with the short shield with rostrum and anteriorly raised margins in accord with anterodorsal shifting of the antennae, the absence of compound eyes, and specialization of the first thoracopods as maxillipeds (Fig. 11F), are remarkable similarities with Copepoda. On the other hand, these features clearly contrast the morphology of Orstenocarida, and that of the members of the thecostracan core. Unique to Skaracarida is the lack of post-maxillipedal trunk limbs, while the absence of a mandibular gnathobase in one of the skaracarid species is probably secondary.

The inclusion of Orstenocarida and Skaracarida within the Maxillopoda necessitates only minor revision of previous diagnoses. A problem is still the absence of gonopores in both Upper Cambrian fossil groups. In our view there is no doubt that the skaracarid material were adults, while there is still some uncertainty for the recognition of *Bredocaris* as such, although this interpretation is favored. Alternatively, it is possible that special structures for reproduction, such as gonopores and gonopods, developed later during evolution of the Maxillopoda, in accord with further specialization of the reproductive system.

Considering the differences between the two fossil maxillopodan orders, it seems not unlikely to us that, besides the

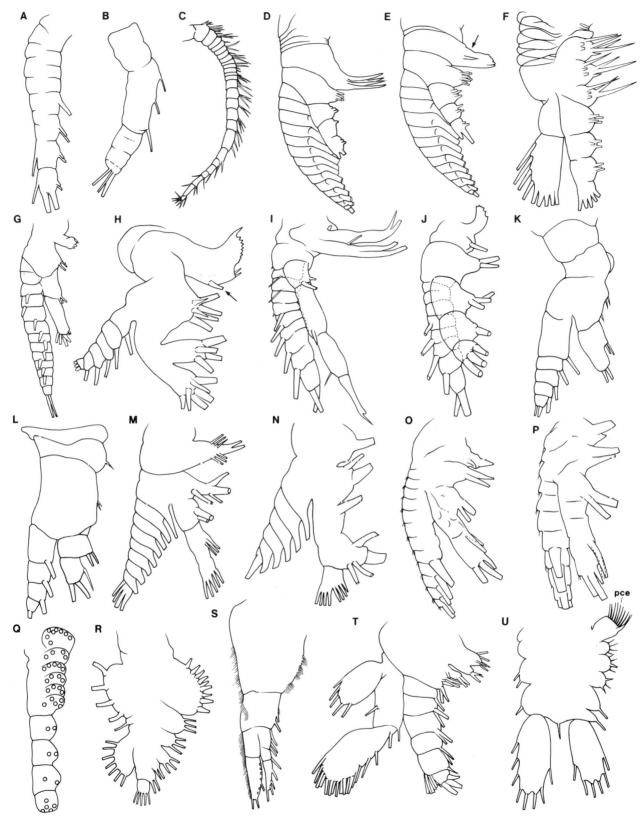

Fig. 11. Selected appendages of Crustacea, most setae shortened, arrows point to gnathobasic setae of mandibular coxae. □ A, D–F. Appendages of the Upper Cambrian skaracarid *Skara minuta* (after Müller & Walossek 1985, Fig. 7A–D): A 1st antenna, D 2nd antenna, E, mandible, F 1st maxilla. □ B. First antenna of nauplius stage 1 of *Neanthessius renicolis* (after Izawa 1986, Fig. 23D). □ C. Male 1st antenna of the calanoid copepod *Eurytemora affinis* (after Katona 1971, Fig. 91). □ G–H, T. Appendages of the cephalocarid *Hutchinssonia macracantha*, G, H 1st larval stage, 2nd antenna and mandible, T 2nd maxilla (after Sanders 1963, Figs. 15B, 35, 37). □ I–J. Second antenna and mandible of first larval stage of the mystacocarid *Derocheilocaris typicus* (Hessler & Sanders 1966, Fig. 3C, D). □ K–L. Second antenna and mandible of 1st nauplius of *Neanthessius renicolis* (Izawa 1986, Fig. 24A, F). □ M–N. Second antenna and mandible of the 6th nauplius of the cirriped *Balanus eburneus* (after Costlow & Bookhout 1957, Fig. 2C–D). □ O, P. Naupliar 2nd antenna and mandible of the ascothoracid *Laura dorsalis* (after Grygier 1985b, Fig. 4J–K). □ Q. Second maxilla of *Derocheilocaris angolensis*, drawn only with attachment points of setae (after Hessler 1971, Fig. 2C). □ R. First maxilla of the 1st copepodid of the copepod *Temora longicornis* (Corkett 1967, Fig. 1D). □ S. Fifth thoracopod of the ascothoracid *Isidascus longispinatus* (after Grygier 1984a, Fig. 2J). □ U. Thoracopod (no. 4–6) of the Devonian lipostracan *Lepidocaris rhyniensis* (after Scourfield 1926, Pl. 23:9).

thecostracan core, there exists another cluster of maxillopodans. This group may be linked by the tendency to reduction of the shield, to formation of a cephalothorax, and
possession of maxillipeds, a cephalo-maxillipedal feeding
apparatus, and a large series of instars. It may include the
Copepoda, the Mystacocarida, and probably the Skaracarida near to its base. Dahl (personal communication) and
Fryer (1985) mentioned the remarkable similarity between
the latter two groups, both having a reduced number of
trunk limbs.

This implies that the level of a six-segmented thorax may
have been reached more than once within the Maxillopoda.
Again, the body segmentation of the Skaracarida (formula
5+6+5+telson), shared with the Tantulocarida, is regarded
as independent retention of the primary condition in both
groups. Another consequence of the redescription is that the
trunk-limb-like 2nd maxilla of the Cephalocarida is not
unique among the Crustacea, but the cephalon with two
pairs of specialized maxillae, as found in most Recent Crustacea, may have been derived independently in the different
subclasses.

Besides the phosphatocopine ostracodes, *Dala peilertae*
(Müller 1983, pp. 94–97) has been described from the same
zone in which *Bredocaris* occurs. *Dala* is one of the largest
orsten crustaceans, with a body length of about 2.2 mm. Its
anterior cephalic region is unknown. The eight-segmented
thorax with a filter apparatus composed of large thoracopods with numerous setiferous endites and a shallow median
food groove, and the five-segmented abdomen composed of
conical rings, the last with branched oar-shaped furcal rami,
were assumed to indicate tentative affinities with Cephalocarida (Müller 1981a). However, review of the type material
and additional specimens recovered since indicates that also
in *Dala* the 2nd maxilla may have been misinterpreted as a
trunk limb, due to its trunk limb-design.

If this be true, the thorax of *Dala* would consist of seven
limb-bearing segments rather than eight, while the head had
five pairs of appendages rather than four. In consequence,
there would be 12 (7+5) trunk segments, including the
separate telson, rather than 13. This would be in accord
with the segmentation of Skaracarida and Tantulocarida,
pointing to close affinities of *Dala* with the Maxillopoda.

Acknowledgements

We thank to Dr. Geoffrey A. Boxshall, London, and Dr.
Robert R. Hessler, La Jolla, for reviewing this paper, various
helpful comments, and linguistic improvements. Valuable
information on special features was provided by Dr. Mark J.
Grygier, Washington, Dr. Tatsunori Ito, Kyoto, and Dr. K.
Izawa, Tsu, Mie Prefecture. D.W. expresses his gratitude to
various colleagues for valuable discussions during a stay at
the Scripps institute of Oceanography, La Jolla, California.

We also appreciate the technical help of Mrs. Andrea
Behr, Mrs. Annemarie Gossmann, Mrs. Dorothea Kranz
(illustrations), Mr. Georg Oleschinski (photography), Mrs.
Helga Rehbach (until 1983), and Mrs. Marianne Vasmer-
Ehses.

The investigations on the Upper Cambrian *orsten* fauna
have been continuously supported by the Deutsche For-
schungsgemeinschaft. Funds for publication have been provided by Statens Naturvetenskapliga Forskningsråd. All examined specimens are deposited in the *Institut für Paläontologie der Rheinischen Friedrich-Wilhelms-Universität Bonn under nos.
UB 640, 641, and UB 882–932.*

References

Andersson, A. 1977: The organ of Bellonci in ostracodes: an ultra-
structural study of the rod-shaped, or frontal, organ. *Acta Zoolo-
gica 58*, 197–204.
Barlow, D.I. & Sleigh, M.A. 1980: The propulsion and use of water
currents for swimming and feeding in larval and adult *Artemia*. *In*
Persoone, G., Sorgeloos, P., Roels, O. & Jaspers, E. (eds.): *The
Brine Shrimp Artemia, Vol. 1. Morphology, Genetics, Radiobiology, Toxi-
cology*, 61–73. Universa Press, Wetteren.
Barnes, M. & Achituv, Y. 1981: The nauplius stages of the cirriped
Tetraclita squamosa rufotincta Pilsbry. *Journal of experimental marine
Biology and Ecology 54*, 149–165.
Barrientos, Y. & Laverack, M.S. 1986: The larval crustacean dorsal
organ and its relationship to the trilobite median tubercle. *Lethaia
19*, 309–313. Oslo.
Behrens, W. 1984: Larvenentwicklung und Metamorphose von *Pyc-
nogonum litorale* (Chelicerata, Pantopoda). *Zoomorphologie 104*, 266–
279.
Benesch, R. 1969: Zur Ontogenie und Morphologie von *Artemia
salina* L.. *Zoologische Jahrbücher der Anatomie 86*, 307–458.
Bowman, T.E. 1971: The case of the nonubiquitous telson and the
fraudulent furca. *Crustaceana 21*, 165–175.
Boxshall, G.A. 1983: A comparative functional analysis of the major
Maxillopodan groups. *In* Schram, F.R. (ed.): *Crustacean Phylogeny*,
121–143. Balkema, Rotterdam.
Boxshall, G.A. 1985: The comparative anatomy of two copepods, a
predatory calanoid and a particle-feeding mormonilloid. *Philo-
sophical Transactions of the Royal Society of London B 311*, 303–377.
Boxshall, G.A., Ferrari, F.D. & Tiemann, H. 1984: The ancestral
copepod, towards a consensus of opinion at the First Internation-
al Conference on Copepoda. *Crustaceana, Suppl. 7, studies on Cope-
poda II*, 68–84. Leiden.
Boxshall, G.A. & Lincoln, R.J. 1983: Tantulocarida, a new class of
Crustacea ectoparasitic on other crustaceans. *Journal of Crustacean
Biology 3(1)*, 1–16.
Boxshall, G.A. & Lincoln, R.J. 1987: The life cycle of the Tantulo-
carida (Crustacea). *Philosophical Transactions of the Royal Society of
London B 315*, 267–303. London.
Brattström, H. 1948: The larval development of the ascothoracid
Ulophysema öresundense Brattström. Studies on *Ulophysema öresun-
dense* 2. *Lunds Universitets Årsskrift. N.F. Avd. 2, 44(5)*, 1–70.
Bresciani, J. 1965: Nauplius 'Y' Hansen. Its distribution and rela-
tion with a new cypris larva. *Videnskabelige Meddelelser fra Dansk
naturhistorisk Forening 128*, 245–259.
Burnett, B.R. 1981: Compound eyes in the cephalocarid crustacean
Hutchinsoniella macracantha. *Journal of Crustacean Biology 1*, 11–15.
Cals, Ph. 1978: Expedition Rumphius II (1975) Crustacés parasites,
commensaux, etc. (Th. Monod et R. Serene, éd.). IV. Crustacés
Isopodes, Gnathiides. Particularités systématiques et morphologi-
ques Appareil piqueur de la larve hématophage. *Bulletin de la
Museum Histoire naturelle de Paris, sér 3, 520, Zoologie 356*, 479–516.
Corkett, C.J. 1967: The copepodid stages of *Temora longicornis* (O.F.
Müller, 1792)(Copepoda). *Crustaceana 12*, 261–273.
Costlow, J.D., Jr. & Bookhout, C.G. 1957: Larval development of
Balanus eburneus in the laboratory. *Biological Bulletin 112*, 313–324.
Costlow, J.D., Jr. & Bookhout, C.G. 1958: Larval development of
Balanus amphitrite var. *denticulata* Broch reared in the laboratory.
Biological Bulletin 114, 284–295.
Dahl, E. 1952: Mystacocarida. Reports of the Lund University
Chile Expedition 1948–49. *Lunds Universitets Årsskrift. N.F. Avd. 2,
48(6)*, 3–41.
Dahl, E. 1956: Some crustacean relationships. *In* Wingstrand, K.G.
(ed.): *Bertil Hanström, Zoological papers in honor of his 65th Birthday*,
138–147. Zoological Institute Lund.
Dahl, E. 1976: Structural plans as functional models exemplified by
the Crustacea Malacostraca. *Zoologica Scripta 6*, 221–228. Oslo.

[Eberhard, C. 1981: Zur Morphologie und Anatomie der vom Naupliusaugenzentrum des Nervensystems der Crustacea innervierten Strukturen. Staatsexamensarbeit, University of Hamburg, 1–157.]

Elofsson, R. 1966: The nauplius eye and frontal organs of the Non-Malacostraca (Crustacea). *Sarsia 25*, 1–128. Bergen.

Fryer, G. 1983: Functional ontogenetic changes in *Branchinecta ferox* (Milne-Edwards)(Crustacea: Anostraca). *Philosophical Transactions of the Royal Society of London 303(1115)*, 229–343.

Fryer, G. 1985: Structure and habits of living branchiopod crustaceans and their bearing on the interpretation of fossil forms. *Transactions of the Royal Society of Edinburgh: Earth sciences 76*, 103–113.

Gauld, D.T. 1959: Swimming and feeding in crustacean larvae: the nauplius larva. *Proceedings of the Zoological Society of London 132*, 31–50.

Gooding, R.U. 1963: *Lightiella incisa* sp. nov. (Cephalocarida) from the West Indies. *Crustaceana 5*, 293–314. Leiden.

Grygier, M.J. 1983a: Ascothoracida and the unity of Maxillopoda. *In* Schram, F.R. (ed.): *Crustacean Phylogeny*, 73–104. Balkema, Rotterdam.

Grygier, M.J. 1983b: A novel, planktonic ascothoracid larva from St. Croix (Crustacea). *Journal of Plankton Research 5(2)*, 197–202. Oxford, England.

Grygier, M.J. 1984a: Ascothoracida (Crustacea: Maxillopoda) parasitic on *Chrysogorgia* (Gorgonacea) in the Pacific and Western Atlantic. *Bulletin of Marine Science 34(1)*, 141–169.

[Grygier, M.J. 1984b: Comparative morphology and ontogeny of the Ascothoracida, a step toward a phylogeny of the Maxillopoda. Ph.D. Thesis, University of California at San Diego, 1–417.]

Grygier, M.J. 1985a: Comparative morphology and ontogeny of the Ascothoracida, a step toward phylogeny of the Maxillopoda. *Dissertation Abstracts International 45(8)*, 2466B–2467B.

Grygier, M.J. 1985b: Lauridae: Taxonomy and morphology of the ascothoracid crustacean parasites of zoanthids. *Bulletin of Marine Science 36(2)*, 278–303.

Grygier, M.J. 1987a: Nauplii, antennular ontogeny, and the position of the Ascothoracida within the Maxillopoda. *Journal of Crustacean Biology 7(1)*, 87–104.

Grygier, M.J. 1987b: Reappraisal of sex determination in the Ascothoracida. *Crustaceana 52(2)*, 149–162.

Grygier, M.J. & Fratt, D.B. 1984: The ascothoracid crustacean *Ascothorax gigas*: redescription, larval development, and notes on its infestation of the Antarctic ophiuroid *Ophionotus victoriae*. *Biology of the Antarctic Seas XVI, Antarctic Research Series 41(2)*, 43–58.

Gurney, R. 1930: The larval stages of the copepod *Longipedia*. *Journal of the Marine Biological Association of the United Kingdom, n.s., 16(2)*, 461–474.

Halcrow, K. & Bousfield, E.L. 1987. Scanning electron microscopy of surface microstructures of some gammaridean amphipod crustaceans. *Journal of Crustacean Biology 7(2)*, 274–287.

Hallberg, E., Elofsson, R. & Grygier, M.J. 1985: An ascothoracid compound eye (Crustacea). *Sarsia 70*, 167–171. Bergen.

Harrington, H.J. 1959: General description of Trilobita. *In* Moore, R.C. (ed.): *Treatise on Invertebrate Paleontology, Part O, Arthropoda 1*, O38–O117. Geological Society of America & University of Kansas Press, Lawrence.

Hartmann, G. 1967: Ostracoda. *In: Dr. H.G. Bronn's Klassen und Ordnungen des Tierreichs, Bd. V, Arthropoda, Abt. 1, Crustacea 2, Teil 4, 2. Lieferung*, 217–408. Geest & Portig, Leipzig.

Hessler, R.R. 1964: The Cephalocarida, Comparative skeletomusculature. *Memoirs of the Connecticut Academy of Arts & Science 16*, 1–96.

Hessler, R.R. 1971: New species of Mystacocarida from Africa. *Crustaceana 21*, 259–273. Leiden.

Hessler, R.R. 1982: 5. Evolution within the Crustacea. Part 1: General: Remipedia, Branchiopoda, and Malacostraca. *In* Abele, L.G. (ed.): *The Biology of Crustacea, vol. 1. Systematics, the Fossil Record, and Biogeography*, 150–185. Academic Press, New York & London.

Hessler, R.R. 1985: Swimming in Crustacea. *Transactions of the Royal Society of Edinburgh: Earth Sciences 76(2)*, 115–122.

Hessler, R.R. & Newman, W.A. 1975: A trilobitomorph origin for the Crustacea. *Fossils and Strata 4*, 437–459, Oslo.

Hessler, R.R. & Sanders, H.L. 1966: *Derocheilocaris typicus* Pennak & Zinn (Mystacocarida) revisited. *Crustaceana 11*, 142–155. Leiden.

Hessler, R.R. & Sanders, H.L. 1971: Two new species of *Sandersiella* (Cephalocarida), including one from the deep sea. *Crustaceana 24*, 181–196. Leiden.

Hicks, G.R.F. 1970: Some littoral harpacticoid copepods, including five new species, from Wellington, New Zealand. *New Zealand Journal of Marine and Freshwater Research 5(1)*, 86–119.

Humes, A.G. 1984: *Hemicyclops columnaris* sp.n. (Copepoda, Poecilostomatoida, Clausiliidae) associated with a coral in Panama (Pacific side). *Zoologica Scripta 13(1)*, 33–39.

Ito, T. 1985: Contributions to the knowledge of Cypris Y (Crustacea: Maxillopoda) with reference to a new genus and three new species from Japan. *Special Publications of the Mukaishima Marine Biological Station*, 113–122.

Ito, T. 1986a: Three types of 'Nauplius Y' (Maxillopoda, Facetotecta) from the North Pacific. *Publications of the Seto Marine Biological Laboratory 31(1/2)*, 63–73.

Ito, T. 1986b: A new species of 'Cypris Y' (Crustacea, Maxillopoda) from the North Pacific. *Publications of the Seto Marine Biological Laboratory 31(3/6)*, 333–339.

Ito, T. 1986c: Problems in the taxonomy of Nauplius Y and Cypris Y (Crustacea). *Zoological Science 3 (TS14)*, IIII.

Izawa, K. 1973: On the development of parasitic Copepoda I. *Sarcotaces pacificus* Komai (Cyclopoida, Philichthyidae). *Publications of the Seto Marine Biological Laboratory 21(2)*, 77–86. Seto.

Izawa, K. 1975: On the development of parasitic Copepoda II. *Colobomatus pupa* Izawa (Cyclopoida, Philichthyidae). *Publications of the Seto Marine Biological Laboratory 22(1/4)*, 147–155. Seto.

Izawa, K. 1986: On the development of parasitic Copepoda IV. Ten species of poecilostome cyclopoids, belonging to Taeniacanthidae, Tegobomolochidae, Lichomolgidae, Philoblennidae, Myicolidae, and Chondracanthidae. *Publications of the Seto Marine Biological Laboratory 31(3/6)*, 81–162. Seto.

Kaestner, A. 1967: *Lehrbuch der Speziellen Zoologie, Bd. I, Wirbellose, 2. Crustacea, 2nd ed.*, 847–1242. Fischer, Stuttgart.

Katona, S.K. 1971: The developmental stages of *Eurytemora affinis* (Poppe 1880) (Copepoda, Calanoida) raised in laboratory cultures, including a comparison with the larvae of *Eurytemora americana* Williams, 1906, and Eurytemora herdmani Thompson & Scott, 1897. *Crustaceana 21*, 5–20. Leiden.

Knox, G.A. & Fenwick, G.D. 1977: *Chiltoniella elongata* n. gen. et sp. (Crustacea, Cephalocarida) from New Zealand. *Journal of the Royal Society of New Zealand 7(4)*, 425–432.

Koehl, M.A.R. & Strickler, J.R. 1981: Copepod feeding currents: food capture at low Reynolds number. *Limnology and Oceanography 26(6)*, 1062–1073.

Kunz, H. 1974: Harpacticoiden (Crustacea, Copepoda) aus dem Küstengrundwasser der französischen Mittelmeerküste. *Zoologica Scripta 3*, 257–282.

Land, M.F. 1984: Crustacea. *In* Ali, M.A. (ed.): *Photoreception and vision in Invertebrates*, 401–438. Plenum Publishing Corporation.

Lauterbach, K.-E., 1973: Schlüsselereignisse in der Evolution der Stammgruppe der Euarthropoda. *Zoologische Beiträge N.F. 19*, 251–299.

Lauterbach, K.-E., 1974: Über die Herkunft des Carapax der Crustaceen. *Zoologische Beiträge N.F. 20(2)*, 273–327.

Lauterbach, K.-E. 1980: Schlüsselereignisse in der Evolution des Grundbauplans der Mandibulata (Arthropoda). *Abhandlungen des Naturwissenschaftlichen Vereins in Hamburg, (N.F.) 23*, 105–161.

Lauterbach, K.-E. 1983: Zum Problem der Monophylie der Crustacea. *Verhandlungen des Naturwissenschaftlichen Vereins in Hamburg, (N.F.) 26*, 293–320.

Lauterbach, K.-E. 1986: Zum Grundplan der Crustacea. *Verhandlungen des Naturwissenschaftlichen Vereins in Hamburg, (N.F.) 28*, 27–63.

Laverack, M.S. & Barrientos, Y. 1985: Sensory and other superficial structures in living marine crustaceans. *Transactions of the Royal Society of Edinburgh: Earth Sciences 76*, 123–136.

Lombardi, J. & Ruppert, E.E. 1982: Functional morphology of locomotion in *Derocheilocaris typica*. *Zoomorphology 100*, 1–10.

Marcotte, B.M. 1977: An Introduction to the Architecture and Kinematics of Harpacticoid (Copepoda) Feeding: *Tisbe furcata* (Baird, 1837). *Mikrofauna Meeresboden 61*, 183–196.

Mauchline, J. 1971: Euphausiacea larvae. *Conseil International pour l'Exploration de la Mer, Zooplankton Sheet 135/137*, 1–16.

Mauchline, J. 1977: The integumental sensilla and glands of pelagic Crustacea. *Journal of the Marine Biological Association of the United Kingdom 57*, 973–994.

McMurrich, J.P. 1917: Notes on some Crustacean forms occurring in the Plankton of Passamaquody Bay. *Proceedings of the Royal Society of Canada, ser. 3, 11, section IV.*, 47–61.

Müller, K.J. 1979: Phosphatocopine ostracodes with preserved appendages from the Upper Cambrian of Sweden. *Lethaia 12*, 1–27.

Müller, K.J. 1981a: Arthropods with phosphatized soft parts from the Upper Cambrian 'Orsten' of Sweden. *Short papers for the Second International Symposium on the Cambrian System. Open-file Report 81-743*, 147–151.

Müller, K.J. 1981b: Softparts of fossils from the Paleozoic Era. *Reports of the DFG – German Research 2/81*, 14–15.

Müller, K.J. 1982a: Weichteile von Fossilien aus dem Erdaltertum. *Naturwissenschaften 69*, 249–254.

Müller, K.J. 1982b: *Hesslandona unisulcata* sp. nov. (Ostracoda) with phosphatized appendages from Upper Cambrian 'Orsten' of Sweden. *In* Bate, R.H., Robinson, E. & Shepard, L. (eds.): *A research manual of fossil and recent ostracodes*, 276–307. Ellis Horwood, Chichester.

Müller, K.J. 1983: Crustacea with preserved soft parts from the Upper Cambrian of Sweden. *Lethaia 16*, 93–109. Oslo.

Müller, K.J. & Walossek, D. 1985: Skaracarida, a new order of Crustacea from the Upper Cambrian of Västergötland, Sweden. *Fossils and Strata 17*, 1–65, Oslo.

Müller, K.J. & Walossek, D. 1986a: *Martinssonia elongata* gen. et sp.n., a crustacean-like euarthropod from the Upper Cambrian 'Orsten' of Sweden. *Zoologica Scripta 15(1)*, 73–92. Oslo.

Müller, K.J. & Walossek, D. 1986b: Arthropodal larval stages from the Upper Cambrian 'Orsten' of Sweden. *Transactions of the Royal Society of Edinburgh: Earth Sciences 77*, 157–179.

Müller, K.J. & Walossek, D. 1986c: Fossils with preserved soft integument as indicators for a flocculent sedimental zone. *12th International Sedimentological Congress, Abstracts*, 221. Canberra, Australia.

Müller, K.J. & Walossek, D. 1987: Morphology, ontogeny, and life habit of *Agnostus pisiformis* from the Upper Cambrian of Sweden. Fossils and Strata 19, 1–124, Oslo.

Newman, W.A. 1982: 5. Evolution within the Crustacea. Part 3: Cirripedia. *In* Abele, L.G. (ed.): *The Biology of Crustacea, vol. 1. Systematics, the Fossil Record, and Biogeography*, 197–220. Academic Press, New York & London.

Newman, W.A. 1983: Origin of the Maxillopoda; urmalacostracan ontogeny and progenesis. *In* Schram, F.R. (ed.): *Crustacean Phylogeny*, 105–119. Balkema, Rotterdam.

Newman, W.A. & Knight, M.A. 1984: The carapace and crustacean evolution – a rebuttal. *Journal of Crustacean Biology 4(4)*, 682–687.

Noodt, W. 1974: Anpassungen an interstitielle Bedingungen: ein Faktor in der Evolution Höherer Taxa der Crustacea? *Faunistisch-Ökologische Mitteilungen 4*, 445–452.

Onbé, T. 1984: The developmental stages of *Longipedia americana* (Copepoda: Harpacticoida) reared in the laboratory. *Journal of Crustacean Biology 4(4)*, 615–631.

Paulus, H.F. 1979: Eye structure and the Monophyly of Arthropoda. *In* Gupta, A.P. (ed.): *Arthropod Phylogeny*, 299–383. Van Nostrand Reinhold Co., New York.

[Perryman, J.C. 1961: The functional morphology of the skeletomusculature system of the larval and adult stages of the copepod *Calanus*, together with an account of the changes undergone by this system during larval development. Ph.D. thesis, University of London, 1–97.]

Rieder, N., Abaffi, P., Hauf, A., Lindel, M. & Weishäupl, H. 1984: Funktionsmorphologische Untersuchungen an den Conchostracen *Leptestheria dahalacensis* und *Limnadia lenticularis* (Crustacea, Phyllopoda, Conchostraca). *Zoologische Beiträge N.F. 28(3)*, 417–444.

Sanders, H.L. 1963: The Cephalocarida. Functional morphology, larval development, comparative external anatomy. *Memoirs of the Connecticut Academy of Arts & Science 15*, 1–80.

Sanders, H.L. & Hessler, R.R. 1964: The larval development of *Lightiella incisa* Gooding. *Crustaceana 7*, 81–97. Leiden.

Schminke, H.K. 1976: The ubiquitous telson and the deceptive furca. *Crustaceana 30*, 293–300. Leiden.

Schminke, H.K. 1981: Adaptation of Bathynellacea (Crustacea, Syncarida) to life in the interstitial ('Zoea Theory'). *Internationale Revue der gesamten Hydrobiologie 66(4)*, 575–637.

Schram, F.R. 1982: 4. The Fossil Record and Evolution of Crustacea. *In* Abele, L.G. (ed.): *The Biology of Crustacea, vol. 1. Systematics, the Fossil Record, and Biogeography*, 94–147. Academic Press, New York & London.

Schram, F.R., Yager, J. & Emerson, M.J. 1986. Remipedia. Part I. Systematics. *San Diego Society of Natural History, Memoir 15*, 1–60. San Diego.

Schram, T.A. 1970a: Marine biological investigations in the Bahamas. 14. Cypris Y, a later developmental stage of nauplius Y Hansen. *Sarsia 44*, 9–24. Bergen.

Schram, T.A. 1970b: On the enigmatic larva nauplius Y type I Hansen. *Sarsia 45*, 53–68. Bergen.

Schram, T.A. 1972: Further records of nauplius Y type IV Hansen from Scandinavian waters. *Sarsia 50*, 1–24. Bergen.

Schrehardt, A. 1986: Der Salinenkrebs Artemia. 2. Die postembryonale Entwicklung. *Mikrokosmos 75(11)*, 334–340.

[Schulz, K. 1976: Das Chitinskelett der Podocopida (Ostracoda, Crustacea) und die Frage der Metamerie dieser Gruppe. Thesis University of Hamburg, 1–167].

Scourfield, D.J. 1926: On a new type of crustacean from the Old Red Sandstone (Rhynie Chert Bed, Aberdeenshire) – *Lepidocaris rhyniensis* gen. et sp. nov. *Philosophical Transactions of the Royal Society of London B 214*, 153–187.

Scourfield, D.J. 1940: Two new and nearly complete specimens of young stages of the Devonian fossil crustacean *Lepidocaris rhyniensis*. *Proceedings of the Linnaean Society London 152*, 290–298.

Shimura, S. 1981: The larval development of *Argulus coregoni* Thorell (Crustacea: Branchiura). *Journal of Natural History 15*, 331–348.

Siewing, R. 1963. Zum Problem der Arthropodenkopfsegmentierung. *Zoologischer Anzeiger 170*, 429–468.

Siewing, R. (ed.) 1985: *Lehrbuch der Zoologie, II, Systematik, 3rd edn.*, 1–1107. Fischer, Stuttgart.

Stubbings, H.G. 1975: *Balanus balanoides. L.M.B.C. Memoirs on Typical British Marine Plants and Animals 37*, 70–171.

Tiemann, H. 1984: Is the taxon Harpacticoida a monophyletic one? *Crustaceana 7*, 47–59.

Tokioka, T. 1936: Larval development and metamorphosis of *Argulus japonicus*. *Memoirs of the College of Science, Kyoto Imperial University. ser. B, 12(1 art. 4)*, 93–114.

Wagin, V.L. 1946: *Ascothorax ophioctenis* and the position of Ascothoracida Wagin in the System of the Entomostraca. *Acta Zoologica 27*, 155–267.

Wagin, V.L. 1954: [On the structure, larval development and metamorphosis of dendrogasterids (parasitic crustaceans of the order Ascothoracida)]. *Uchenye Zapiski Leningradskogo Ordena Lenina Gosudarstvennogo Universiteta Imeni A.A. Zhdanova, Seriya Biologicheskaya 35 (172)*, 42–89.

Walley, L.J. 1969: Studies on the larval structure and metamorphosis of *Balanus balanoides* (L.). *Philosophical Transactions of the Royal Society of London 256, B 807*, 237–280.

Walossek, D. & Müller, K.J., in press: A second type A-nauplius from the Upper Cambrian 'Orsten' of Sweden. *Lethaia*.

Weigmann-Haass, R. 1977: Die Calyptopis- und Furcilia-Stadien von *Euphausia hanseni* (Crustacea: Euphausiacea). *Helgoländer wissenschaftliche Meeresuntersuchungen 29*, 315–327.

Yager, J. 1981: Remipedia, a new Class of Crustacea from a marine Cave in the Bahamas. *Journal of Crustacean Biology 1*, 328–333.

Yager, J. & Schram, F.D. 1986: *Lasionectes entrichoma*, new genus, new species, (Crustacea: Remipedia) from anchialine caves in the turks and caicos, British West Indies. *Proceedings of the Biological Society of Washington 99(1)*, 65–70.

Addendum

Two additional papers of Grygier appeared after this manuscript was submitted. Grygier 1987c presents new data on facetotectan larvae and also discusses various characters of this group in respect of the copepods, branchiurans, ascothoracids, cirripeds, and faceto-

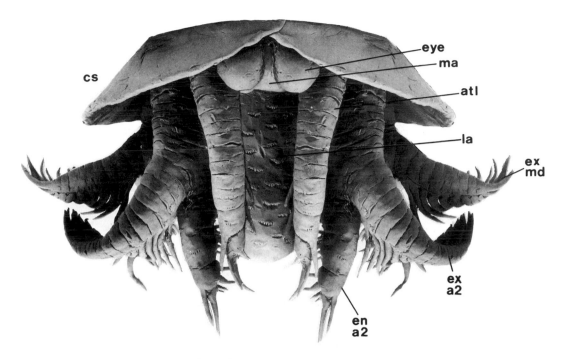

Fig. 12. Plasticine model of adult *Bredocaris*. Frontal view; left side completed photographically.

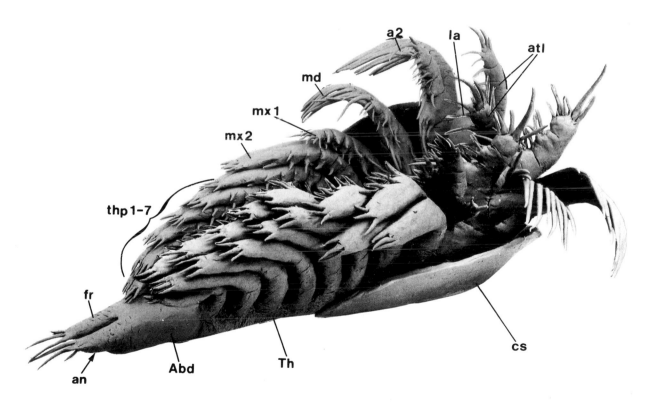

Fig. 13. Plasticine model of adult *Bredocaris*. Ventrolateral view, slightly from posterior; model completed as in Fig. 12.

tectans. His table with 'cladistically useful characters' and the phylogram based on these data may be questioned in detail. However, they show an interesting agreement with our study: the copepods are considered to be the most plesiomorphic and opposite to all other groups investigated, belonging to the thecostracan core in our paper.

Grygier 1987d is a description of a new ascothoracid larva from the Antarctic. It shows the subsequent stage under the cuticle having eight pairs of limb buds on its trunk, the 2nd maxillae, six pairs of thoracopods, and probably buds of the genital limbs, which we regard as modified thoracopods.

Grygier, M.J. 1987c: New records, external and internal anatomy, and systematic position of Hansen's Y-larvae (Crustacea: Maxillopoda: Facetotecta). *Sarsia 72*, 261–278. Bergen.

Grygier, M.J. 1987d: Antarctic records of asteroid-infesting Ascothoracida (Crustacea), including a new genus of Ctenosculidae. *Proceedings of the Biological Society of Washington 100(4)*, 700–712.

Plate 6

Posterior trunk region and surface structures.

☐ 1. Same specimen as in Pls. 1:3, 4; 4:5–9; 5:1, 2; 15:3; 16:3. Almost dorsal view of posterior end of trunk, with posterior four thoracopods of left side, abdomen and furcal rami; note the different surface textures of thorax and abdomen.

☐ 2. Same specimen as in Pl. 1:6; 6:5. Ventral surface of abdomen (Abd) immediately behind last thoracic sternite, showing a group of short transverse furrows (compare with similar structures in larval specimens on Pls. 9:7; 15:7; see p. 11).

☐ 3. Same specimen as in Pls. 2:1; 3:2. View of abdomen, continuing into the furcal rami (fr); supraanal flap (arrow) recognizable below dorsocaudal spine (dscp; cs = cephalic shield; thp = thoracopods).

☐ 4. UB 896 (same specimen as in Pl. 6:7). Posterolateral view of abdomen; due to slight inflation, the abdominal surface shows numerous fine folds, and the supraanal flap (saf) is protruding; rows of denticles (den) give the impression of a weak segmentation of the furcal rami; the four terminal setae (s) are broken off distally (an = anus; dcsp = dorsocaudal spine).

☐ 5. Same specimen as in Pl. 1:6; 6:2. Dorsal view of abdomen and furcal rami. Fragments of setae on left ramus.

☐ 6. UB 897. Posteroventral view of abdomen and furcal rami; left ramus entirely preserved, with denticles (arrows) and proximal parts of the terminal setae (thp7 = last thoracopod).

☐ 7. Same specimen as in Pl. 6:4. Ventral view of abdomen and rami; it is unknown whether the groove (arrow) is indicative of a particular structure, since it could not be recognized on other specimens.

8–9. Same specimen as in Pl. 2:6.

☐ 8. View of wrinkled and twisted furcal rami, with remnants of all terminal setae: two thick ones insert medially and two thinner ones laterally, above the other.

☐ 9. Surface texture of thorax, probably enhanced by slight shrinkage of the body wall; there is no distinct outer segmentation on dorsal and lateral sides of the fleshy trunk (but see ventral side on Pl. 5:6); on lower side: bases of the thoracopods.

☐ 10. Same specimen as in Pls. 2:9; 5:5; 16:4. Detail of the ventral cuticle of abdomen and right furcal ramus, covered with delicate furrows and denticles.

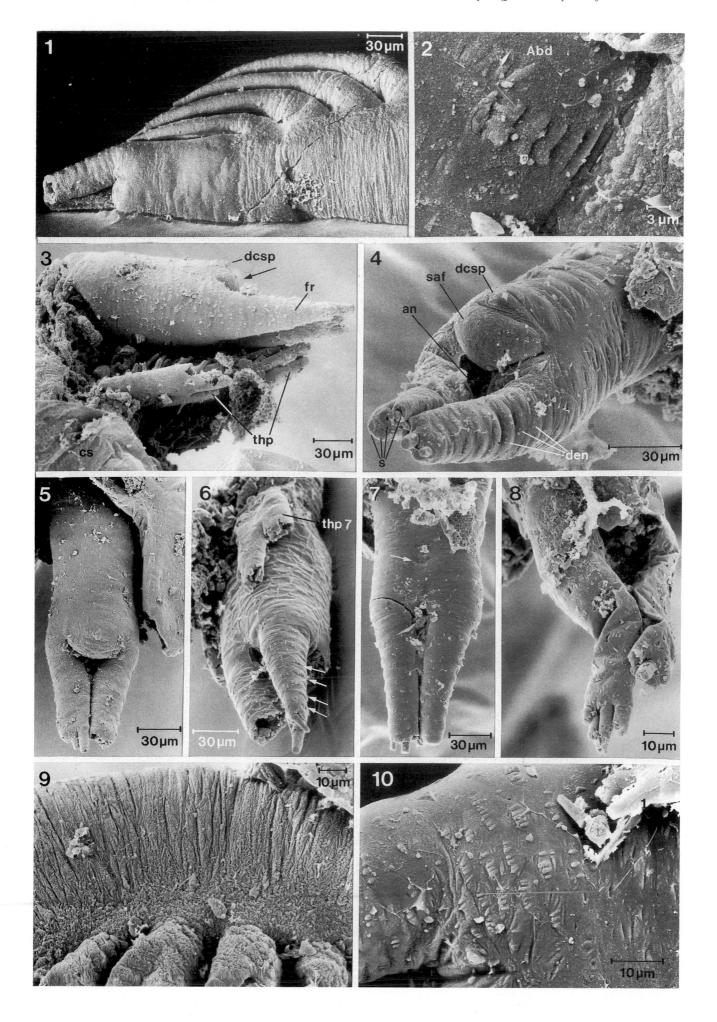

Plate 7

Instar I.

☐ 1. UB 898 (same specimen as in Pls. 7:3, 8; 8:1, 3, 6–8). Subventral view of complete specimen; rudimentary 1st maxilla (rud mx1) still with setae terminally; initial furcal rami (i fr) terminating in one thick seta and a thinner, subordinate one laterally; prominent dorsocaudal spine (dcsp) broken off distally; some of the denticles and setules on various parts of the body indicated by arrows; note the robust enditic spines of the 2nd antenna and mandible at the side of the labrum; during scanning parts of the appendages were broken off (compare, e.g., 8 with Pl. 8:1; atl = 1st antenna; a2 = 2nd antenna; gn = mandibular gnathobase; la = labrum; md = mandible; st = sternite of mandibular segment).

☐ 2. UB 899 (same specimen as in Pl. 7:6). Ventral view; cephalic shield (cs) slightly deformed; anterior three cephalic appendages (atl, a2, md) only poorly preserved, except the mandibular coxae; eye lobes (lo) visible in front of the prominent and inflated labrum (la); 1st maxilla (rud mx1) lying close on the larval trunk; one of the furcal setae (s) is preserved with almost its entire length (dcsp = dorsocaudal spine).

☐ 3. Same specimen as in Pls. 7:1, 8; 8:1, 3, 6–8. Almost lateral view; cuticle finely wrinkled, indicating slight shrinkage prior to fossilization; cephalic shield (cs) only weakly defined; some of the exopodal setae of 2nd antenna and mandible (ex a2, md) preserved but broken off distally (photo taken prior to breakage).

☐ 4. UB 900. Ventral view; anterior body region somewhat distorted; eye (eye) preserved but wrinkled; labrum pressed posteriorly, broken off distally; 1st antennae broken off; most of left 2nd antenna (a2), mandibular coxae (md), and right exopod present but much crumpled.

☐ 5. UB 901 (same specimen as in Pl. 8:2, 5, 10). Ventral view of almost complete, collapsed specimen; accordingly, all appendages (atl, a2) are posteriorly directed, lying one above the other and thus disguising the mandibles; some of their setae are still present; eye (eye) and labrum (la) well-preserved; 1st maxilla (rud mx1) still with a seta on its lateral peak, the initial exopod (i ex); groups of denticles (den) cover the anterior cuticle of the labrum (cs = cephalic shield; i fr = initial furcal rami).

☐ 6. Same specimen as in Pl. 7:2. Lateral view; due to slight compression, the lateral margin of the cephalic shield (cs) is better defined than in 3; shield terminating clearly behind 1st maxilla; dorsocaudal spine almost complete (arrow; eye = probable compound eye).

☐ 7. UB 902 (same specimen as in Pl. 8:11). Lateral view. Specimen mounted onto its distorted anterior head portion; hind body well-preserved, with 1st maxilla, furcal rami, and dorsocaudal spine (la = wrinkled labrum).

☐ 8. Same specimen as in Pls. 7:1, 3; 8:1, 3, 6–8. Frontal view; eye lobes are separated by apparently softer area, extending from anterior shield margin to labral base; ventrally pointing appendages almost complete, some of the setae still present; arrow points to the weakly defined shield margin (most of shield not visible due to mounting; photo taken prior to breakage, see Pl. 8:1).

☐ 9. UB 903. Lateral view of incompletely and coarsely preserved specimen (recognizable in particular by the small holes on the trunk); labrum preserved, pointing posteroventrally; mandibular rami (md) and rudimentary 1st maxilla, initial furcal rami and dorsocaudal spine still present (cs = cephalic shield).

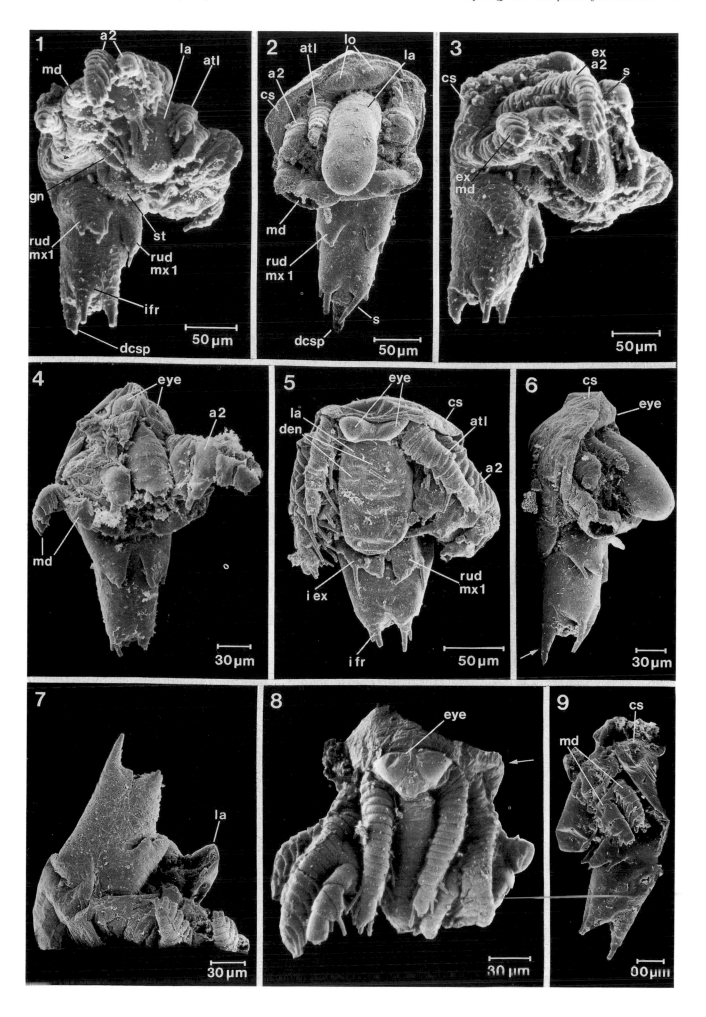

Plate 9

Instar II.

☐ 1. UB 906 (same specimen as in Pl. 10:4, 5). Posteroventral view; mandibular coxae (md) well-preserved; 1st maxilla (mx1) developed, flanking the larval trunk laterally (an = anus; a2 = 2nd antenna; gn = gnathobase; la = labrum).

☐ 2. UB 907 (same specimen as in Pl. 11:6). Lateral view; cephalic shield and trunk with dorsocaudal spine (dcsp) well-preserved, but labrum and appendages rather incomplete; eye (eye) anteroposteriorly compressed (mx1 = 1st maxilla).

☐ 3. UB 908 (same specimen as in Pls. 9:6; 10:2; 11:5). Lateral view; cephalic shield somewhat deformed but well-defined, with slightly overhanging margins; anterior three cephalic appendages preserved only with their proximal parts; 1st maxilla posteriorly directed, with segmented endopod (en) and short one-segmented exopod (ex); dorsocaudal spine (dcsp) broken off distally (eye = probable compound eye).

☐ 4. UB 909. Ventrolateral view of well-inflated, slightly deformed trunk fragment; furcal rami (fr) complete except the terminal setae (s); denticles (den) on posterior end of trunk and furcal rami, some of them appear to be arranged as scales; arrows point to limb buds of 2nd maxilla and first thoracopod.

☐ 5. UB 910 (same specimen as in Pls. 10:1; 11:4). Ventral view of wrinkled but almost complete specimen; due to collapsing the 1st maxilla (mx1) in particular is deeply withdrawn into the body.

☐ 6. Same specimen as in Pls. 9:3; 10:2; 11:5. Dorsal view; cephalic shield oval-shaped, widest and highest in the first third (between 2nd antenna and mandible) and narrowing posteriorly; trunk conically tapering towards the basis of the dorsocaudal spine (do = plate-like area on highest point of the shield; mx1 = 1st maxilla).

☐ 7. UB 911 (same specimen as in Pl. 10:3). Ventral view of laterally compressed specimen; due to collapsing all ventral structures are withdrawn into the cephalic shield (cs); eye well-preserved, also the midventral area (ma); fragment of right 2nd antenna (a2), mandibular coxae (md) and 1st maxillae (mx1) preserved, labrum (la) distorted; anlagen of 2nd maxilla (mx2) and first thoracopod (thp1) recognizable on the trunk; arrow points to a group of small furrows on the trunk.

8–9. UB 912 (same specimen as in Pls. 10:3, 6, 7; 11: 1–3).

☐ 8. Ventral view; cephalic shield slightly deformed and ventral structures slightly laterally dislocated; eye (eye) preserved, but midventral area collapsed; labrum and right set of appendages almost absent, left set complete, except the endopods of 2nd antenna and mandible (atl = 1st antenna; a2 = 2nd antenna; an = anus; cox = mandibular coxa; fr = furcal rami; md = mandible; mx1 = 1st maxilla; tlb = trunk limb buds).

☐ 9. Lateral view; cephalic shield clearly enclosing more than the maxillulary segment; shield somewhat deformed; appendages exceptionally well-preserved; note the similarity in shape of the anterior three appendages.

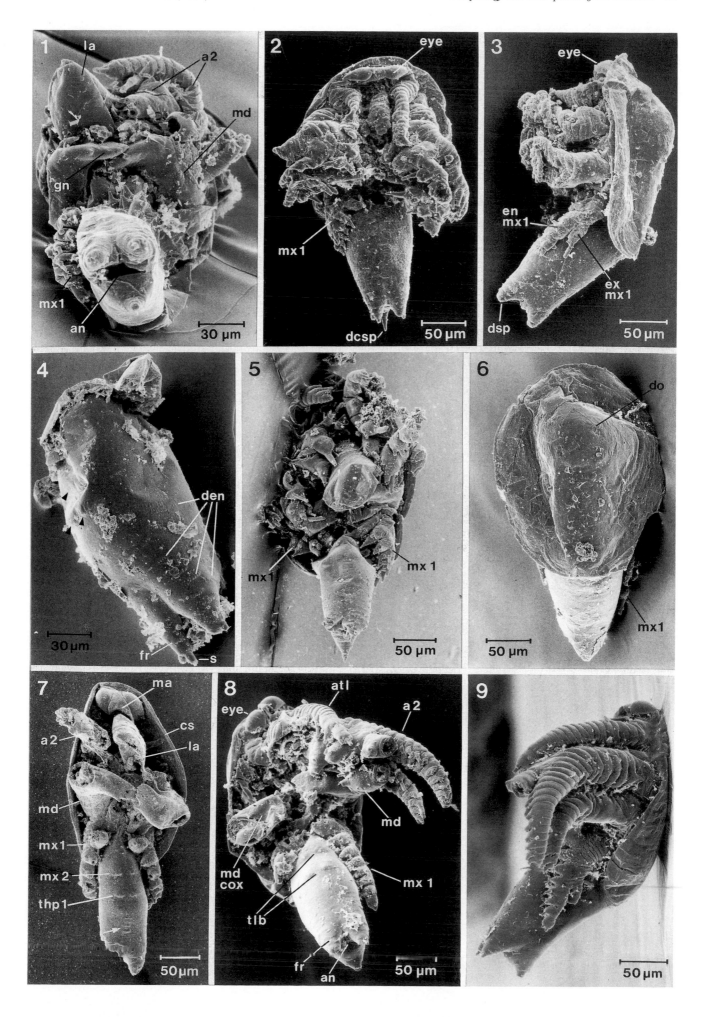

Plate 10

Details of instar II.

□ 1. Same specimen as in Pls. 9:5; 11:4. Dorsal view of distorted cephalic shield; arrows point to pores at posterior margin of apical plate (compare with Pls. 3:2; 9:6); on upper left: 1st maxilla.

□ 2. Same specimen as in Pls. 9:3, 6; 11:5. Anterior view; anterior margin of cephalic shield slightly raised medially, exposing the eye (eye); labrum (la) deformed, being flanked by the 1st antennae (atl); appendages rather coarsely phosphatized; note the sharp edge of the shield (a2 = 2nd antenna; md = mandible).

□ 3. Same specimen as in Pls. 9:8, 9; 10:6, 7; 11:1–3. Close-up of probable compound eye; midventral area laterally compressed; in lower left corner: basis of right 1st antenna.

4–5. Same specimen as in Pl. 9:7.

□ 4. Close-up of eye similar to 3, but midventral area better preserved, posterior margin of lobes appears to be slightly tipped; a window indicates the area shown in 5 (alp = alien particles).

□ 5. Detail of anterior region of the midventral area with a knob- or pore-like structure (compare with Pls. 8:2; 14:8).

6–7. Same specimen as in Pls. 9:8, 9; 10:3; 11:1–3.

□ 6. Anterior view of complete left antennae (atl, a2); latter appendage lacking the distal podomeres of its endopod (en); rows on distal margins of ringlets well-preserved (bas = basipod; mx1 = 1st maxilla).

□ 7. Distal end of 1st antenna; setae broken off distally; arrow points to the small terminal podomere forming the basis of midmost two setae; note the fine holes on the surface, indicating slightly incomplete phosphatization.

□ 8. UB 913. Close-up of mandible, showing the articulation between prominent mandibular coxa (cox) and wrinkled distal limb portion (arrow; bas = basipod; cs = cephalic shield; en = endopod; ex = exopod).

□ 9. Same specimen as in Pl. 9:7. Mandibular coxae with blade-like, medially pointing gnathobases (distal end not preserved); marginal spinules of different size, a larger one (arrow) on the posterior margin, somewhat set off from the row on inner margin; gnathobasic setae broken off (gns; compare with Pl. 4:4; m = hole probably indicative of the mouth).

□ 10. UB 914. Distorted mandibular coxal gnathobase with almost complete distal seta; window shows area of 11.

□ 11. Same specimen. Enlarged view of gnathobasic seta, originally bearing axial rows of thin setules (some indicated by arrows).

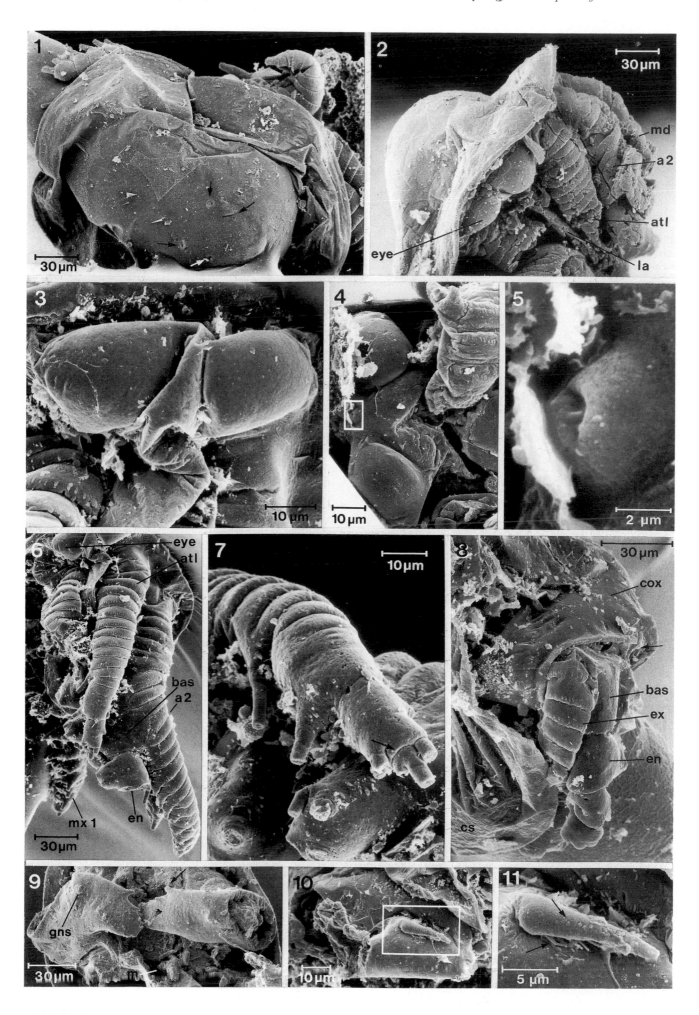

Plate 11

1–8: instar II.

9, 10: instar III.

1–3. Same specimen as in Pls. 9:8, 9; 10:3, 6, 7.

☐ 1. Median view of posteriorly directed 1st maxilla, flanking the hind body (see also Fig. 5B); on left: anlagen of 2nd maxilla and first thoracopod (pce = proximal endite).

☐ 2. Anteromedian view of 1st maxilla; at least on the proximal six enditic protrusions, bases of posterior seta or spine slightly drawn out (pce = proximal endite).

☐ 3. Anterior view. Segmentation of maxillulary protopod is unclear: from this photograph it is likely that the proximal endite (pce) belongs to the shaft (sh), which is finely folded on its outer edge; the next one or two endites may belong to the coxa (cox?), while fourth segment is the basipod (bas?), which carries the two rami (en, ex; see pp. 12–13).

☐ 4. Same specimen as in Pls. 9:5; 10:1. Anteromedian view of 1st maxilla; proximal portion somewhat depressed; some of the enditic spines preserved with their entire lengths; distal endopodal podomere with pair of setae, as in the more proximal podomeres, and a further, apical seta; exopod covered by the shield.

☐ 5. Same specimen as in Pls. 9:3, 6; 10:2. Coarsely preserved 1st maxilla; terminal seta of short exopod still present; arrows point to sharp-edged fractures of the surface which indicate a secondary layer of phosphate, causing a thickening of surface structures (secondary coating, see also Müller & Walossek 1985, p. 6, 7, Fig. 2).

☐ 6. Same specimen as in Pl. 9:2. Ventral view of the larval trunk, slightly from posterior; as the 1st maxilla is positioned on the anterior body portion in this stage, the larval trunk starts with the maxillary segment; limb buds of 2nd maxilla and first thoracopod only faintly recognizable, second one being only a minute hump (tlb).

☐ 7. UB 915. Close-up of limb buds; anterior pair slightly further developed and posteriorly projecting, its distal margin is faintly fringed (arrows), indicating the future development of endites (compare with Pls. 9:7 and 13:10).

☐ 8. UB 916. View of crumpled trunk, showing arrangement and varying sizes of denticles on initial furcal rami (i fr) and base of dorsocaudal spine (dcsp).

9–10. UB 917.

☐ 9. Subventral view of fragmentarily preserved specimen; three pairs of limb buds (tlb) are developed on the hind body: 2nd maxilla and two pairs of thoracopods; anterior two buds slightly more developed than in preceding instar (compare with 6, 7 and Pl. 9:7; eye = probable compound eye; la = remains of labrum; md = distorted mandibular coxae).

☐ 10. Enlarged view of the posterior two limb buds on the ventrally sloping trunk surface (arrow).

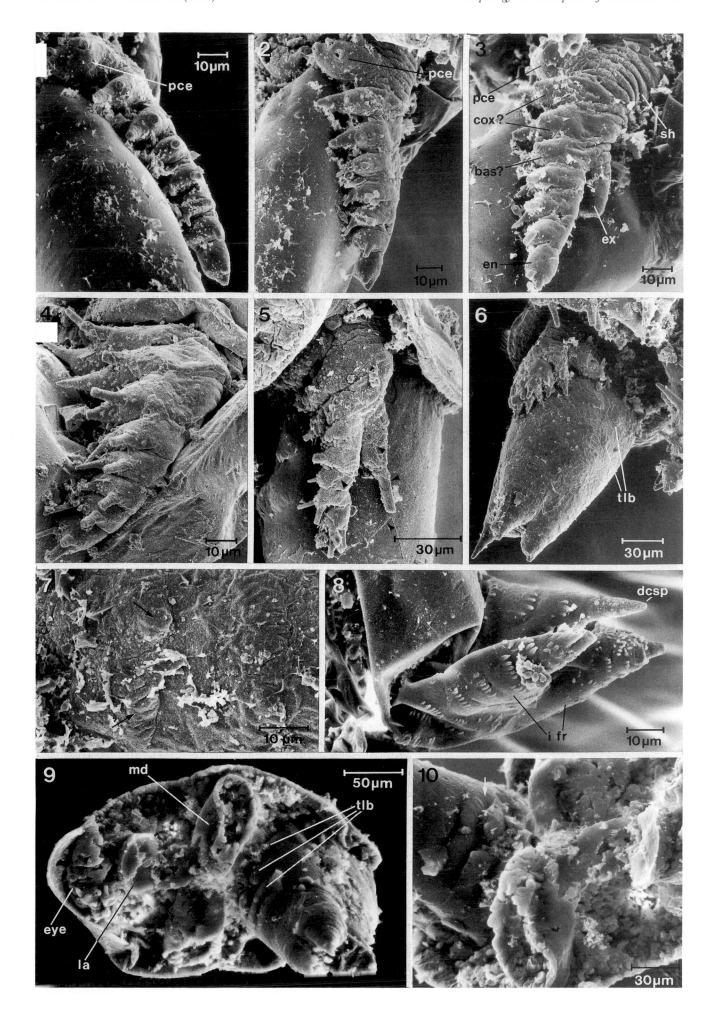

Plate 12

Instar IV.

1, 2, 4–8: UB 918 (same specimen as in Pl. 13:1, 2, 4, 5, 8, 9, 11).

□ 1. Lateral view; cephalic shield slightly deformed anteriorly; appendages partly preserved (a2, md, mx1); trunk well-preserved and inflated, ventrally flexed (fr = furcal rami; la = labrum).

□ 2. Anterior view; midventral area between eye lobes (lo) covered with alien particles; 1st antennae not preserved; only distorted proximal parts of 2nd antennae and exopod of left one preserved (ex a2); labrum (la) ventrally projecting; furcal rami (fr) longer than in preceding stages (compare with Pls. 7; 9; 11:9; md = mandible).

□ 3. UB 919. Ventral view of anteriorly distorted specimen; breakaway of labrum (la) permits observation of mandibular gnathobase (gn) and 1st maxilla (mx1); four pairs of limb buds are developed on the trunk, which gradually decrease in size and state of development; anterior buds bilobate, indicating the two rami (arrow).

□ 4. Ventral view of exceptionally well preserved specimen; shield widely extending from the ventral body, posterior margin slightly concave (arrow); trunk ventrally flexed; post-antennular appendages (a2, md, mx1) ventrodistally directed and arranged around the postoral food chamber, bordered anteriorly by the ventrally projecting labrum anteriorly and the trunk posteriorly; strong mandibular coxal gnathobases (gn) point medially and approach each other, endites of antennal coxa and basipod (cox, bas) flank the

sides of the labrum; strong enditic spines (esp) of mandibular basipod preserved in part; maxillulary endites anteromedially directed (alp = alien particles; an = anus; il = inner lamella; tlb = limb buds).

□ 5. Ventrolateral view, showing the orientation of the appendages, labrum and trunk; weak segmentation of caudal end of abdomen and furcal rami caused by arrangement of denticles; dorsocaudal spine broken off distal to its insertion (arrow).

□ 6. View of labrum from almost posterior direction; four to five rows of setules are developed along cuticular ridges at the side of the labrum (arrows; md gn = medially overlapping mandibular gnathobases).

□ 7. Ventral view of postoral food chamber, being bordered by the labrum (la) and by the trunk (tr); due to the anterior direction of the mandibular coxae (md cox), their gnathobases are almost vertically oriented, closing the atrium oris; some of the setae on the proximal endites of 1st maxillae (pce) preserved in part; sternal surface covered with alien particles (window indicates area of 8); note the angle between coxal body and gnathobase.

□ 8. Close-up of anteriorly pointing distal surface of the left gnathobase, covered with numerous setules which appear to be arranged in rows parallel to the inner margin; inner margin with spinules of varying size, a larger spinule more posteriorly positioned (arrow; la = labrum).

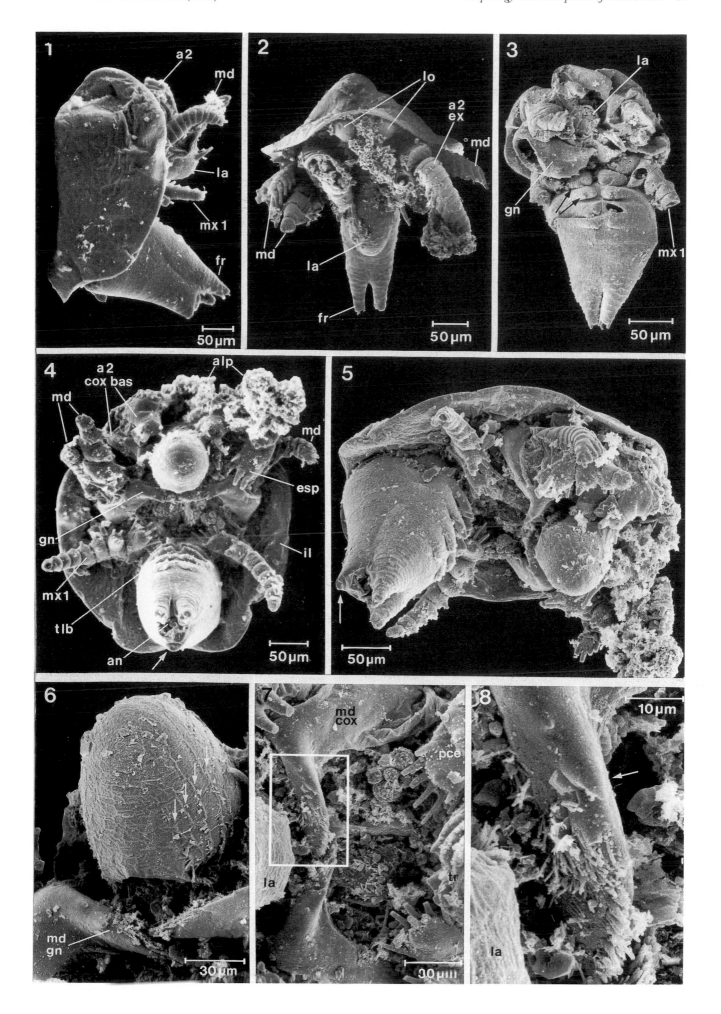

Plate 13

Details of instar IV.

1, 2, 4, 5, 8, 9, 11: same specimen as in Pl. 12:1, 2, 4–8.

3, 6, 10: UB 920 (same specimen as in Pl. 14:2).

☐ 1. Median view of right mandible, slightly from posterior; endopod complete except the setation; exopod (ex) posteriorly flexed due to collapsing (compare with Pls. 1:3; 4:2; 7:4; 9:8); arrows point to some of the setules originally covering the median surface of the limb (a2 = 2nd antenna; cs = cephalic shield; la = labrum).

☐ 2. Almost posterior view of left mandible; well-sclerotized and uniform coxa (cox) with oblique anteriorly angled gnathobase (gn); basipod (bas) articulating on the distal coxal surface and exopod (ex) crumpled; enditic spines partly preserved (esp; en = endopod).

☐ 3. Slightly deformed mandibular coxa and gnathobase, the distal surface of which is covered with numerous fine setules; gnathobasic seta (gns) indicated only by its insertion point (bas = basipod).

☐ 4. Posterior view of 1st maxilla, showing the horn-like produced posteromedian sides of the endites; exopod (ex) arising from outer margin of fourth protopodal segment with two terminal setae; distal segment of endopod (en) slightly distorted (pce = proximal endite).

☐ 5. View of median surface of complete 1st maxilla (compare with Pls. 4:9; 10:6, 7; 14:5); proximal endite somewhat flattened; note the successive change in shape of the endites and in number and arrangement of setae from proximal to distal; distal segment with a row of setae apically; entire median surface of the limb originally furnished with fine setules (tlb = limb buds on hind body).

☐ 6. View of the flat anterior side of 1st maxilla; proximal endite and distal two endopodal podomeres not preserved; from this particular specimen it may also be possible that the basipod was composed of two elements rather than one, while the coxa was uniform and bore only one endite (compare with Pls. 4:7, 9; 11:1–5; 18:8; see pp. 12, 13).

☐ 7. UB 921. Close-up of proximal endite of 1st maxilla, showing setules on the enditic surface and double rows of subordinate setules on the marginal setae (= pectinate setae); direction of setules different due to position of the setae around the margin.

☐ 8. View of proximal two maxillulary endites from posteromedially; posterior side of second endite produced into horn-like extension, continuing into a stout spine and another short one at its base.

☐ 9. View of the four pairs of lobate limb buds on the trunk; anterior two buds are distinctly bifid distally, posterior buds are only triangular lobes; size decreasing from anterior (on right side) to posterior.

☐ 10. Enlarged view of some of the anterior limb buds; inner rims with shallow indentations (arrows), indicating the future division into endites there; lobes with two pointed humps of different sizes terminally: larger one represents the future endopod, smaller outer one represents the future exopod; trunk surface finely wrinkled, most probably due to slight shrinkage prior to fossilization.

☐ 11. Lateral view of trunk; furcal rami with short rows of denticles; terminal setae and dorsocaudal spine not preserved; trunk surface slightly shrunken, as in 10.

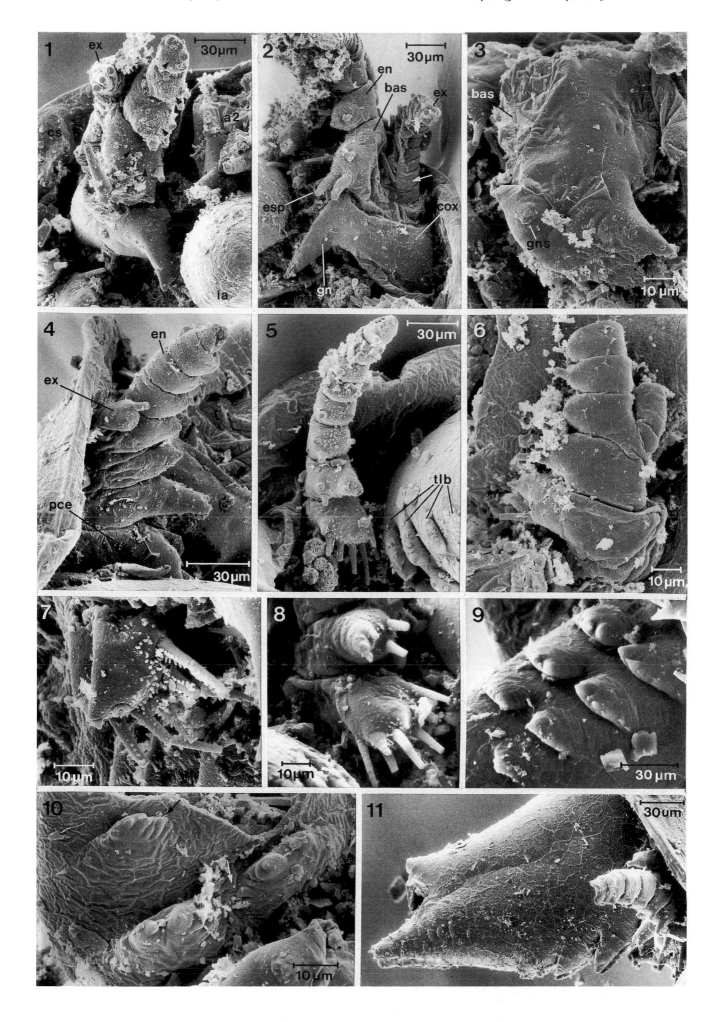

Plate 14

1, 2: instar IV.

3–8: instar V.

☐ 1. UB 922. Posterior view of larval trunk with dorsocaudal spine (dcsp), anus (an) and furcal rami (fr); rami with three terminal setae (cs = cephalic shield).

☐ 2. Same specimen as in Pl. 13:3, 6, 10. View from posterior on posterior of cephalic shield (cs); furrow (fu) running from behind 1st maxilla (mx1) towards the posterior margin of the shield represents the tagma border between larval head and trunk; right and left furrows do not join at the margin (arrow; compare with Pl. 3:7, 8); inner lamella inflated (il), probably due to gas production within the body cavity prior to fossilization; trunk (tr) finely wrinkled anteriorly, similar to the presumed adult.

☐ 3. UB 923 (same specimen as in Pls. 14:8; 15:1, 2, 6, 8). Ventrolateral view; cephalic shield sunken into glue of adhesive tape; 1st antennae broken off; post-antennular limbs (a2, md, mx1) wrinkled but rather complete; right 2nd antenna anteriorly stretched, mandible and 1st maxilla posterolaterally directed; limb buds (tlb) on ventrally bent larval trunk covered with alien particles (la = crumpled labrum).

☐ 4. UB 924 (same specimen as in Pl. 15:4). Dorsal view of incomplete and coarsely preserved specimen (shield with numerous holes); 1st antennae (atl) and endopod of right 2nd antenna (en a2) far anteriorly stretched.

☐ 5. UB 925. Ventral view of anteriorly distorted specimen; ventral structures of cephalon destroyed, cephalic shield dislocated and heavily crumpled; trunk well-preserved, with 5 pairs of limb buds and furcal rami; buds progressively decreasing in size and degree of development.

☐ 6. UB 926. Ventral view of similarly preserved specimen; few more of the ventral cephalic structures preserved, covered by coarse alien particles; left 1st maxilla (mx1) complete and posteriorly stretched; incomplete phosphatization caused numerous holes in the surface of the trunk and limb anlagen.

☐ 7. UB 927. Lateral view of specimen with distorted cephalon and complete but slightly laterally compressed trunk; appendages fragmentarily preserved (md, mx1; dcsp = dorsocaudal spine; tlb = trunk limb buds).

☐ 8. Same specimen as in Pls. 14:3; 15:1, 2, 6, 8. Ventral view of probable compound eye (compare with Pls. 3:3–6; 7:2, 5, 8; 8:1; 2; 9:7; 10:2–5; 12:2); lobes are slightly posteriorly projecting; arrow points to the pore- or pimple-like structure on the deformed midventral area (atl = 1st antenna).

Plate 15

Details of instar V and addendum to adult morphology.

1, 2, 6, 8. Same specimen as in Pl. 14:2, 8.

□ 1. Anterior (distal) surface of labrum, furnished with short rows of denticles (compare with Pl. 7:4, 5).

□ 2. Ventral view of postoral food chamber; due to deformation of the labrum (la), the lateral rows of setules are located in a fold on the posterior side (arrow); orally oriented mandibular gnathobases (gn) covered with crystallites; gnathobases approaching each other medially; sternum (st) furnished with thin setules, which are broken off distally (gns = gnathobasic seta; tr = trunk).

□ 3. Same specimen as in Pls. 1:3, 4; 4:5–9; 5:1, 2; 6:1; 16:3. Close-up of maxillulary exopod; terminal setae preserved in part; denticles or setules on distal end of ramus, some also on one of the setae.

□ 4. Same specimen as in Pl. 14:4. Distal podomeres of 1st antenna (at1) and endopod of 2nd antenna (en a2); proximal parts of setae preserved; rows of denticles may indicate a former subdivision of the distal podomeres; note the similarity between these two limbs; arrows point to small distal podomeres of antennula and antennal endopod.

□ 5. UB 928. Anterior view of posteriorly bent 2nd antenna; coxa and shaft not visible; elongate basipodal endite (end bas) still with spines (esp); basipod with two ringlets on the outer surface; proximal endopodal podomere endopod also produced into an endite; proximal exopodal ringlets lacking setae, arrows point to setae not corresponding to the annuli (compare with Pls. 4:6; 8:6; 9:9; 10:6; la = labrum).

□ 6. View of median surfaces of mandible (md) and 1st maxilla (mx1); both limbs and the sternal surface (st) are covered with numerous setules, which are partly disguised by alien particles (gns = gnathobasic seta of mandibular coxa; tr = larval trunk).

□ 7. UB 929. Ventrolateral view of limb buds; size and shape of buds decreases progressively towards the posterior; anterior buds somewhat incompletely phosphatized distally; last bud little pronounced; between the latter pair and the furcal rami (fr) few short transverse furrows can be seen (compare with Pls. 6:2; 9:7; 11:6).

□ 8. Anterolateral view of mandible (md) and 1st maxilla (mx1); on the mandibular exopod each seta corresponds to one annulus (see 5 for exopod of 2nd antenna); limbs and trunk covered with numerous alien particles; on lower right: large articulation between mandibular coxa and basipod.

□ 9. UB 930. Ventral view of postoral region back to trunk limb buds; specimen roughly and somewhat incompletely preserved; larval head and trunk are separated by a deep transverse furrow; sternites of mandibular (md) and maxillulary (mx1) segments fused to a single plate; two elevations (arrows) arising behind the mandibular coxae (cox) represent the initial paragnaths (compare with Pl. 5:6); on lower left corner: mouth opening (m).

□ 10. UB 931. Posteroventral view of incomplete trunk, the interior of which is filled with a conical mass; it is, however, unknown whether this filling refers exclusively to the gut; trunk limb buds of this specimen distinctly bifid.

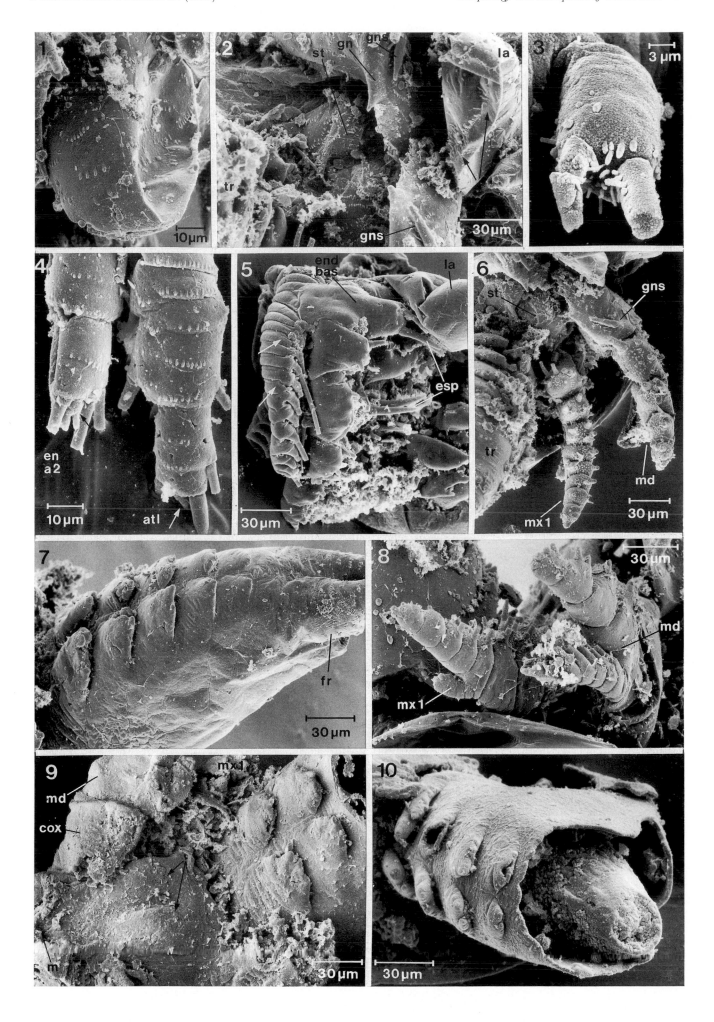

Plate 16

Thoracopods.

☐ 1. UB 889 (same specimen as in Pl. 2:3). Ventral view of trunk fragment with complete 4th thoracopod, breakage of 3rd one permits view of the cross-section of the protopod; proximal endites (pce) anteriorly directed.

☐ 2. UB 886 (same specimen as in Pls. 1:9; 4:4; 5:3, 4). Lateral view of 5th and 6th thoracopods; distal portions flexed posteriorly at proximal border of basipod (bas, arrow).

☐ 3. UB 640 (same specimen as in Pls. 1:3, 4; 4:5–9; 5:1, 2; 6:1; 15:3). View of anterior surfaces of the complete 4th to 6th thoracopods; proximal limb portion finely wrinkled (arrow as in 2; bas = basipod; cox = coxa).

☐ 4. UB 894 (same specimen as in Pls. 2:9; 5:5; 6:10). Sixth and 7th thoracopods; arrows indicate borders of basipod (bas = basipod; cox = coxa; en = endopod; ex = exopod).

5–8 UB 932.

☐ 5. Posterior view of one of the anterior thoracopods, dislocated after breakage; limb almost complete except for the endopod (en) and the exopodal (ex) setae; median endites are progressively more distally oriented; note the oval articulation area of the hollow basis of the limb (bas = basipod; pce = proximal endite).

☐ 6. Almost lateral view, slightly from posterior; position shows the flatness of the limb (sh = shaft).

☐ 7. Median view, almost from the limb base, showing the different orientation of the triangular proximal endite (pce) in relation to the more distal endites.

☐ 8. View of the flattened posterior side of the limb, almost from the tip of limb; from proximal to distal the enditic spines become progressively more widely spaced.

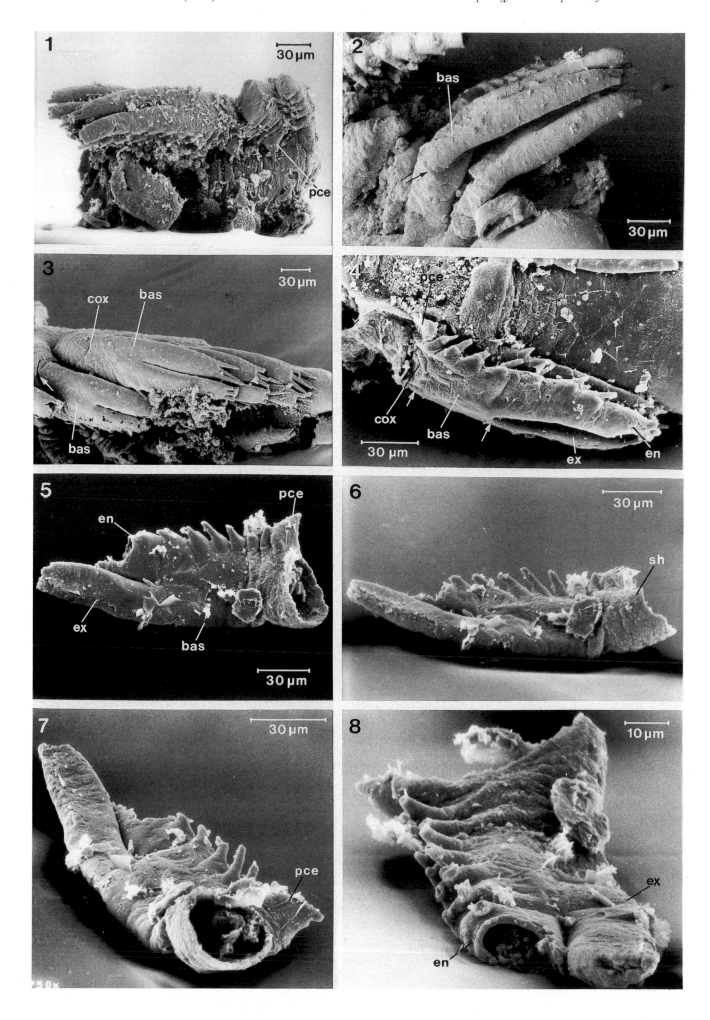

List of abbreviations

Abd	abdomen
Ad	presumed adult
alp	alien particle
an	anus
app	appendages
atl	first antenna, antennula
a2	second antenna, antenna
bas	basipod
C	cephalic region, cephalon
cox	coxa
cs	cephalic shield
dcsp	dorsocaudal spine
den	denticles
do	'dorsal organ', plate-like structure on cephalic shield, with two pairs of pores at posterior margin
en	endopod
end	enditic process, endite
esp	enditic spine
ex	exopod
eye	probable compound eye
ff	frontal filaments of cirriped and facetotectan nauplii
fr	furcal ramus
fu	furrow
g	gut, digestive tract
gn	gnathobase, grinding plate of mandibular coxa
gns	gnathobasic seta, positioned on flattened distal surface of mandibular gnathobase
h	height
hol	holotype
i	initial, incipient
il	inner lamella, cuticle below shield
l	length

la	labrum
lo	lobes of probable compound eye
loc	locality
L1–5	larval stages
m	mouth
ma	area between lobes of probable compound eye and labrum
md	mandible
mx1	first maxilla, maxillula
mx2	second maxilla, maxilla
ne	naupliar eye
par	paratype
pce	proximal endite of post-mandibular appendages
pgn	paragnaths, pair of outgrowths on sternite of mandibular segment
prot	protopod
rud	rudimentary, of larval shape
s	seta
saf	supraanal flap, operculum
sh	shaft
spec	specimen
st	sternite, sternal bar
stl	setule, delicate thin bristle
T	trunk region of adults (= thorax + abdomen)
Th	thoracic region
thp1–7	thoracopods
tl	total length
tlb	trunk limb bud
tr	larval trunk, hind body
UB	repository number
w	width
wi	window, translucent cuticular area on dorsal shield of facetotectan nauplius, above naupliar eye